Mohammad Hossein Keshavarz
Combustible Organic Materials

Also of Interest

The Properties of Energetic Materials.
Sensitivity, Physical and Thermodynamic Properties
Keshavarz, Klapötke, 2017
ISBN 978-3-11-052187-0, e-ISBN 978-3-11-052188-7

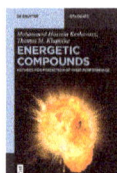

Energetic Compounds.
Methods for Prediction of their Performance
Keshavarz, Klapötke, 2017
ISBN 978-3-11-052184-9, e-ISBN 978-3-11-052186-3

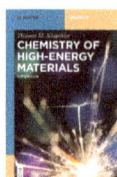

Chemistry of High-Energy Materials.
4th Edition
Klapötke, 2017
ISBN 978-3-11-053631-7, e-ISBN 978-3-11-053651-5

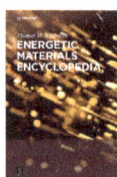

Energetic Materials Encyclopedia.
Klapötke, 2018
ISBN 978-3-11-044139-0, e-ISBN 978-3-11-044292-2

Mohammad Hossein Keshavarz

Combustible Organic Materials

Determination and Prediction of Combustion Properties

DE GRUYTER

Author
Prof. Dr. Mohammad Hossein Keshavarz
Malek-ashtar University of Technology
Department of Chemistry
83145 115 Shahin-Shahr
Iran
Keshavarz7@gmail.com

ISBN 978-3-11-057220-9
e-ISBN (PDF) 978-3-11-057222-3
e-ISBN (EPUB) 978-3-11-057233-9

Library of Congress Control Number: 2018008913

Bibliographic information published by the Deutsche Nationalbibliothek
The Deutsche Nationalbibliothek lists this publication in the Deutsche Nationalbibliografie; detailed bibliographic data are available on the Internet at http://dnb.dnb.de.

© 2018 Walter de Gruyter GmbH, Berlin/Boston
Cover image: art-skvortsova/iStock/Getty Images Plus
Typesetting: Integra Software Services Pvt. Ltd
Printing and binding: CPI books GmbH, Leck

www.degruyter.com

Preface

For a combustible organic material, it is important to assess its combustion properties. This book introduces four chapters about flash point, autoignition, flammability
limit, and heat of combustion of different classes of combustible organic compounds.
Chapter 5 also demonstrates heat release, specific heat release rate, and heat release
capacity as three important characteristics of polymer flammability. Each chapter
discusses suitable methods for experimental determination of the mentioned combustion properties. Different predictive methods that are in current use are also
reviewed. Among diverse approaches, simple and reliable ones are discussed with
some examples. There is no need to use computer codes, unfamiliar descriptors, and
expert users in the methods illustrated in the chapters. Furthermore, each chapter
contains some complementary problems with their answers, which can improve the
expertise of students and researchers when applying the models for predicting the
desired combustion property of a newly designed or synthesized organic compound.

https://doi.org/10.1515/9783110572223-201

Contents

1 Flash Point

Flash point (FP) of an organic liquid compound is the lowest temperature at which a liquid can form an ignitable mixture in air near the surface of the liquid where its measurement requires an ignition source. The American Society for Testing and Materials (ASTM) defines FP under specified testing conditions and 101.3 kPa pressure [1]. The FP data are widely used to evaluate the fire and explosion hazards of liquids. They have great practical significance in the handling and transporting of such chemicals in bulk quantities. Above the FP temperature, a liquid can produce enough vapor to form a flammable mixture with air. Organic liquids with lower FP can be ignited more easily because the FP of a volatile organic compound is the lowest temperature at which vapors of a fluid will ignite. Fire point is the temperature at which the vapor continues to burn after being ignited. When the ignition source is removed at the FP, which is different from the fire point, the vapor may cease to burn. Since FP and the fire point do not depend on the temperature of the ignition source, a certain concentration of vapor for each flammable liquid in the air is necessary to sustain combustion. Thus, when an ignition source of sufficient strength is applied, the FP of an organic flammable liquid is the lowest temperature at which there will be enough flammable vapor to ignite.

The FP of a pure flammable organic compound or mixture is an important parameter in hazard classification of flammable liquids. Since the FP is an approximation of the lower temperature limit, the temperature at which a chemical compound evolves enough vapors to support combustion, it is essential to be updated for safe handling, transportation, and storage of many substances [2]. The values of FP for common organic compounds are widely reported because the knowledge of the combustion potential of an organic compound is crucial when designing safe chemical processes. They can also help firefighters in extinguishing fires because a high FP liquid fire pool can be extinguished by cooling with water mist. Meanwhile, liquids with low values of the FP usually need to be blanketed by dry chemicals or foams [2, 3]. Since cooking oil, for example, has a high FP and its temperature during burning is high (603–733 K), its gasification heat is also high and additional thermal energy is required for fuel evaporation. As the rate of supply of fuel vapor or burning rate is reduced sufficiently enough not to support the flame, the cooking oil fire can be extinguished by water mist mainly through cooling the fuel surface [4]. Due to the presence of large heat capacity of water, it absorbs high quantities of heat and pulls down the temperature of the fire. Thus, water is a good candidate for extinguishing the fires of high FP liquids. Since foam can act as a blanket and not as a heat absorber, it can be used for extinguishing the fires of high or low FP [2]. For industrial processes, the FP is also used to determine the vapor explosion potential [5, 6].

https://doi.org/10.1515/9783110572223-001

1.1 Measurement of the FP

There are five main apparatuses used internationally for measurement of the FP that are given in Table 1.1. Table 1.2 also shows eight different test methods of ASTM, which has been defined based on four of these apparatuses. As seen in Table 1.2, there are two methods for measuring the FP [3]: (1) the closed cup method and (2) the open cup method, which are used for liquids with low and high FP, respectively. The open cup method can measure the FP using the conditions that are met in open vessels and that would be encountered in spills [7]. The vessel is conducted in a container where it is exposed to the air outside. The temperature of the desired substance in the vessel is gradually raised and an ignition source is passed over the top of it. At the FP, the temperature reaches a point at which it "flashes" and ignites. The recorded FP will vary according to the height of the source above the cup, i.e., the distance between the substance and the ignition source. Since low boiling components of the mixture may be lost to the surrounding atmosphere prior to the application of the flame, which gives higher values of the FP [7], this is one disadvantage of the open tester. Closed cup methods may give somewhat lower results than open cup methods. This situation is due to the presence of a physical barrier, which prevents the volatile particles from escaping and causes to approximate an equilibrium between vapor and the air in the

Table 1.1: Five main FP devices currently used.

Device	Temperature uniformity	Sample volume (mL)	Main use
Tagliabue (Tag)	Liquid Bath	50	Less viscous compounds
Pensky-Martens	Stirred; Metal shell	75	Viscous compounds
Cleveland	Metal plate across cup base	70	Open cup tests
Small-Scale (Setaflash)	Preheated to fixed temperature	2 or 4	Small-scale and flash/no-flash tests
Abel	Water bath and air gap	79	European tests

Table 1.2: ASTM standardized method for measuring the FP.

Method	Device	Open/ Closed	Heating rate °C·min^{-1}	Range (°C)
D 56	Tag	Closed	1–3	<93
D 92	Cleveland	Open	5–6	79–400
D 93	Pensky-Martens	Closed	4–5	40–370
D 1310	Tag	Open	1	−18 to 165
D 3278	Small-scale	Closed	Fixed temperature	0–110
D 3828	Small-scale	Closed	Fixed temperature	−20 to 300
D 3941	Tag or Pensky-Martens	Closed	Fixed temperature	0–110
D 3941	Tag or Pensky-Martens	Closed	0.5	<93 or 40–370

enclosed space [3, 7]. The closed cup tester gives an insight into the flammability of materials within enclosed spaces, e.g., sealed containers [7]. Experimental determinations of the FP are described in various sources, which differ in their scope and experimental conditions as:

(a) The ASTM D56 [8] (Tag Closed Cup method, which is shown in Figure 1.1) is used for materials with viscosities lower than 5.5 mm²/s and the FP below 93 °C.

(b) The ASTM D93 [9, 10] (Pensky Martens Closed Cup method, which is shown in Figure 1.2) is used for materials with the FP between 40 °C and 360 °C, such as distillate fuels, lubricating oils, fuel oils, mixtures of petroleum liquids with solids, and biodiesel fuels.

(c) The ASTM D92 [10, 11] (Cleveland Open Cup method, which is shown in Figure 1.3) is used for liquids with high viscosity and the FP between 79 °C and 400 °C, such as petroleum products except for fuel oils.

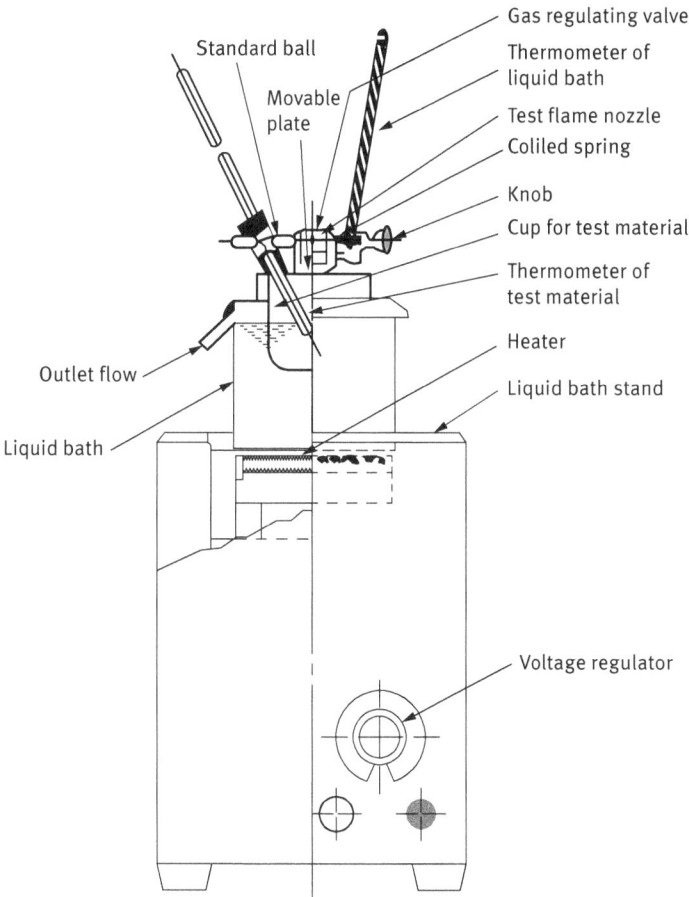

Figure 1.1: Tag closed cup method for measuring the FP.

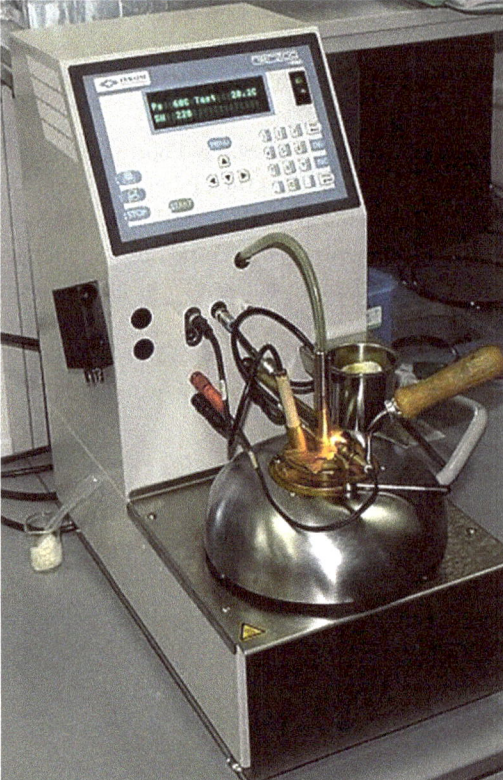

Figure 1.2: Pensky-Martens closed cup tester.

Figure 1.3: Cleveland open cup method.

Based on the type of material and the available apparatus, one can select each of these standard methods. Details of these procedures for the determination of the FP of unknown substances and the equality of the test methods are reviewed elsewhere [10, 12].

Different factors can affect the measured FP values, which include [6, 7, 13–15]:
(1) Sample size as well as its viscosity and homogeneity (in mixtures)
(2) The type, position, and dimension of the ignition source
(3) Temperature rise rate
(4) Stirring of liquid
(5) Mixing vapor phase above pool
(6) Drafts
(7) Ambient pressure
(8) Operator bias
(9) Fuel container condition (open or closed)

Thus, the reported experimental values of FP in literature sometimes differ tens of degrees [12].

1.2 Predictive methods of the FP for pure compounds

Measurement of the FP temperature is costly and may contain high experimental uncertainties [16]. Since the values of the FP are not available in the literature for many organic compounds, it is important to have suitable predictive methods for those organic compounds where their experimental values of the FP were not reported in the literature. Moreover, prediction of the FP is also important for toxic, explosive, or radioactive materials because the measurements of their FP are very difficult and even impossible in some cases [17, 18]. Thus, reliable prediction of the FP values for different classes of organic compounds is desirable.

The lack of the measured data of the FP for most newly introduced compounds in the industry has driven the development of many FP predictive models, which can be generally divided into three categories: empirical, group contribution, and quantitative structure-property relationship (QSPR) models. There are several reviews of these methods [19–21].

1.2.1 Empirical models

Empirical models include correlations, which predict the FP by other physical properties that are more accessible or can be measured more easily, e.g., normal boiling point

(NBP) temperature, density, vapor pressure, critical properties, and enthalpy of vaporization. Prugh [22] introduced a simple technique for predicting the approximate FP of a compound through its atomic composition and boiling point. Fujii and Hermann [23] suggested several correlations between the FPs and vapor pressures at 25 °C for pure organic compounds. They found that classification of organic compounds according to their chemical-structural characteristics (e.g., functional groups) is essential. For pure organic compounds, a linear correlation is the most appropriate correlation to explain the relationship between the inverse of the FP (K) and the logarithm of vapor pressures (mm Hg) at 25 °C [23]. Patil [24] introduced a general correlation for estimation of the FPs of organic compounds from their NBPs where estimated FPs are within a deviation of about 10% of the reported values. Due to less accuracy of this correlation, Patil [24] also recommended separate correlations for estimation of the FPs of acids and alcohols. Satyanarayana and Kakati [25] proposed a general correlation for predicting the FPs of organic compounds from their NBPs and specific gravity, where the predicted FPs are within an average deviation of 8.301 of the reported values. Satyanarayana and Rao [26] proposed an improved correlation in the form a non-linear exponential type for the estimation of the FP of organic compounds and petroleum fractions as a function of their NBPs. Metcalfe and Metcalfe [27] improved and simplified the correlation of Satyanarayana and Kakati [25] where the influence of the chemical functionality on the FPs has been clearly demonstrated. Hshieh [28] developed a correlation between the FP and the NBP for silicone compounds. The NBP has also been used to introduce a more general correlation for organic compounds including silicone compounds. For silicone compounds and general organic compounds (including silicone compounds), the standard errors of the estimate are 11.06 and 11.66 K, respectively [28]. Catoire and Naudet [19] presented a simple empirical equation for estimation of the FP of pure organic liquids. Their correlation is based on NBP, the standard enthalpy of vaporization at 298.15 K [$\Delta_{vap}H°(298.15$ K$)$] of the compound, and the number of carbon atoms (n_C) in the molecule. The bounds for this correlation are $-100 \le FP(°C) \le +200$; $250 \le NBP$ (K) ≤ 650; $20 \le \Delta_{vap}H°(298.15$ K$)/(kJmol^{-1}) \le 110$; and $1 \le n_c \le 21$. The accuracy of these empirical correlations is directly dependent on the accuracy of the measured physical properties and the predictive methods, which were used for their estimations. If one of the aforementioned properties such as the NBP is not available, it is not possible to estimate the FP property. This is the main drawback of empirical correlations.

Among different empirical methods, the procedures based on the NBP are the most widely used methods because the NBP is the most readily available thermophysical properties. It should be mentioned that the value of the FP of a substance is related directly to the NBP and inversely to the vapor pressure at a given temperature. Flammable liquids with high vapor pressures, as a general rule, at "normal" temperatures commonly exhibit low boiling points and low FP [7].

1.2.1.1 The use of NBP

Some theoretical methods have been developed for the calculation/prediction of the FP of pure compounds or mixtures of flammable liquids [20, 21, 29–33]. As mentioned before, Catoire and Naudet [19] introduced a simple correlation for predicting the FP values of pure organic molecules on the basis of their NBPs, $\Delta_{vap}H°(298.15K)$, and n_C. Gharagheizi et al. [34] offered a linear model for estimation of the FP of pure compounds on the basis of their NBPs and chemical structures. For those approaches, which need experimental data of physical and thermodynamic properties of organic compounds for predicting the FP values, e.g., the methods of Catoire and Naudet [19] as well as Gharagheizi et al. [34], they may give large deviations. This situation may be satisfied for those compounds where the required physical and thermodynamic properties have not been given in the literature. For those classes of organic compounds where the models have been developed, these methods can provide good predictions of the FP, if experimental data of desired variables such as the NBPs are available. Due to the ease of measurement of NBP, these methods are highly attractive. Simple models have been proposed in the literature, which includes:
(1) Quadratic [24, 28]
(2) Exponential [26]
(3) Non-linear correlations between the FP and the NBP [35–37]
(4) The other correlations proposed to predict the FP using the NBP and liquid density [25, 27].

The other relationships exist between the NBP and some properties, e.g., (i) n_C[38]; (ii) enthalpy of vaporization [19, 39, 40]; (iii) critical pressure, critical temperature, acentric factor, and molecular weight [41]. Some of these methods are reviewed here.

1.2.1.2 The NBP and enthalpy of vaporization at the NBP

Rowley et al. [39] have introduced a suitable correlation based on the NBP and enthalpy of vaporization at the NBP. The new correlation provides an absolute average deviation of 1.3% when compared with data for more than 1,000 compounds. It has the following form:

$$\begin{aligned} FP = 0.2554 \, \frac{NBP}{(\beta+1)^{0.2164}} \left(\frac{\Delta_{vap}H(\text{at NBP})}{R}\right)^{0.1606} + 6.31(\beta+1)^{0.7} \\ - 5.6 \times 10^{-5} \frac{NBP}{(\beta+1)} \left(\frac{\Delta_{vap}H°(298.15\,\text{K})}{R}\right) + 0.0208 \left(\frac{\Delta_{vap}H°(298.15\,\text{K})}{R}\right)^{0.8767} \end{aligned}$$

(1.1)

where $\Delta_{vap}H$ (at NBP) is the enthalpy of vaporization at the NBP, and R is the gas constant. The parameter β is defined as:

$$\beta = n_C + n_{Si} + n_S + \frac{n_H - n_X - 2n_O}{4} \tag{1.2}$$

where n_C, n_{Si}, n_S, n_H, n_X, and n_O are the number of carbon, silicon, sulfur, hydrogen, halogen, and oxygen atoms in the compound, respectively. When information pertaining to $\Delta_{vap}H°(298.15\,K)$ is not readily available, as is often the case without an accurate vapor pressure correlation, it may be calculated from $\Delta_{vap}H$ (at NBP) and the critical temperature, T_C, using the Watson correlation [42]:

$$\Delta_{vap}H° \,(298.15\,K) = \Delta_{vap}H \,(\text{at NBP}) \left(\frac{1 - \dfrac{298.15\,K}{T_C}}{1 - \dfrac{NBP}{T_C}} \right)$$

Since this substitution results in only a small change in the accuracy of the method, the parameter $\Delta_{vap}H$ (at NBP) may be used instead of $\Delta_{vap}H°(298.15\,K)$ in order to reduce the complexity of the model, i.e., $\Delta_{vap}H = \Delta_{vap}H°(298.15\,K) \approx \Delta_{vap}H$ (at NBP). Thus, eq. (1.1) can be rewritten as:

$$\begin{aligned} FP = 0.2554 \,\frac{NBP}{(\beta+1)^{0.2164}} \left(\frac{\Delta_{vap}H}{R} \right)^{0.1606} + 6.31(\beta+1)^{0.7} \\[2mm] -5.6 \times 10^{-5} \,\frac{NBP}{(\beta+1)} \left(\frac{\Delta_{vap}H}{R} \right) + 0.0208 \left(\frac{\Delta_{vap}H}{R} \right)^{0.8767} \end{aligned} \tag{1.3}$$

Example 1.1: Calculate the FP for *o*-dinitrobenzene using eq. (1.3) where the NBP and $\Delta_{vap}H$ are 592.2 K and 60.0 kJ mol^{-1} [43], respectively.

Answer: The molecular formula of *o*-dinitrobenzene is $C_6H_4N_2O_4$. Thus, the FP is calculated as:

$$\beta = n_C + n_{Si} + n_S + \frac{n_H - n_X - 2n_O}{4} = 6 + 0 + 0 + \frac{4 - 0 - 2 \times 4}{4} = 5$$

$$\begin{aligned} FP = 0.2554 \,\frac{NBP}{(\beta+1)^{0.2164}} \left(\frac{\Delta_{vap}H}{R} \right)^{0.1606} + 6.31(\beta+1)^{0.7} \\[2mm] -5.6 \times 10^{-5} \,\frac{NBP}{(\beta+1)} \left(\frac{\Delta_{vap}H}{R} \right) + 0.0208 \left(\frac{\Delta_{vap}H}{R} \right)^{0.8767} \end{aligned}$$

$$\begin{aligned} FP = 0.2554 \,\frac{592.2}{(5+1)^{0.2164}} \left(\frac{60.0 \times 10^3}{8.314} \right)^{0.1606} + 6.31(5+1)^{0.7} \\[2mm] -5.6 \times 10^{-5} \,\frac{592.2}{(5+1)} \left(\frac{60.0 \times 10^3}{8.314} \right) + 0.0208 \left(\frac{60.0 \times 10^3}{8.314} \right)^{0.8767} = 459.9\,K \end{aligned}$$

Since the experimental data of the FP for this compound is 423 K [43], its deviation is 36.9 K.

1.2.1.3 The NBP and n_C

Gharagheizi et al. [38] presented an empirical method involving the NBP temperature and n_C of the pure compounds for predicting the FP temperature of pure substances. They used a total of 1,471 pure compounds belonging to 77 chemical families to develop a general correlation. The global absolute average deviation of the model that results from experimental values is 2.4%. They obtained the following empirical correlation from a regression of the calculated results over the experimental FP temperature values:

$$FP = -18.44 + 0.8493 \times NBP - 3.723 \times n_C \tag{1.4}$$

Example 1.2: Use eq. (1.4) and calculate the FP for 4-methyl-1,2-dinitrobenzene with the following molecular structure; NBP = 610 K [43]:

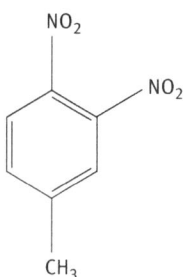

Answer: Since n_C in this compound is seven, the use of eq. (1.4) gives:

$$FP = -18.44 + 0.8493 \times NBP - 3.723 \times n_C$$
$$= -18.44 + 0.8493 \times 610 - 3.723 \times 7 = 473.57 \text{ K}$$

Since the experimental FP of this compound is 453 K [43], its error is 20.57 K.

Gharagheizi et al. [38] compared their model with the output results of Catoire-Naudet [19], Patil [24], and Hshieh [28] where eq. (1.4) gives a more reliable prediction.

1.2.1.4 Linear correlation between the FP, the NBP, and predefined functional groups constituting the molecule

Alibakhshi et al. [44] used experimental data of the FP and the NBP data of 1,533 organic compounds from various chemical families to derive their relationship. They analyzed the chemical structures of dataset compounds for obtaining 42 functional groups, which are listed in Table 1.3. In their study, functional groups were defined such that they can characterize the whole structure of all dataset compounds where they are almost similar to those applied by Albahri [45] and Lazzús [46]. Since prediction of the FP through linear structural group contribution (SGC) method resulted with high errors, they improved the accuracy of linear SGC method by considering the linear correlation between the FP and the NBP, which was also proposed in some works [25, 27, 28, 38]. Alibakhshi et al. [44] introduced the following linear relationship as an accurate model to predict the FP:

$$FP = 12.14 + 0.73NBP + \sum n_i \varphi_i \tag{1.5}$$

where n_i and φ_i are the number and the amount of contribution of functional group i, respectively. Alibakhshi et al. [47] improved the constants of eq. (1.5) for the dataset containing only the experimental data as:

$$FP = 11.07 + 0.72NBP + \sum n_j \varphi_j \tag{1.6}$$

The values of φ_i and φ_j for different groups are given in Table 1.3.

Table 1.3: Functional groups and their contribution to FP and NB.

Functional Group	φ_i	φ_j	Functional Group	φ_i	φ_j
$-CH_3$	−2.97	−0.65	$-C\equiv N$	5.84	7.31
$-CH_2-$	−1.14	−1.34	$-NO_2$	7.41	10.33
$>CH-$	−1.64	−4.11	$-F$	2.84	3.74
$>C<$	−0.1	−5.92	$-Cl$	7.33	9.65
$=CH_2$	−2.08	−0.76	$-Br$	7.66	6.17
$=CH-$	−2.21	−1.52	$-I$	12.17	11.06
$=C<$	−2.03	−3.06	$-SH$	−1.77	0.79
$=C=$	−1.01	−3.69	$-S-$	−0.84	−0.97
$=CH$	−4.36	−2.01	$-CH_2-$(ring)	−2.49	−1.91
$=C-$	−0.23	−1.73	$-HC<$ (ring)	−0.26	−1.57
$-OH$	12.63	14.94	$=CH-$ (ring)	−1.4	−1.11
$-O-$	2.67	3.22	$>C<$ (ring)	−2.52	−6.62
$>C=O$	5.02	3.8	$=C<$ (ring)	−0.39	−1.13
$-CHO$ (aldehyde)	4.92	3.39	$-O-$ (ring)	4.13	8.92
$-COOH$ (acid)	15.33	23.4	$-OH$ (ring)	10.78	12.92
$-COO-$ (ester)	5.33	6.95	$>C=O$	5.38	5.73
$HCOO-$ (formate)	3.86	6.83	$-NH-$ (ring)	8.14	12.61
$-NH_2$	6.91	7.57	$>N-$ (ring)	14.67	0.62
$-NH-$	2.74	5.38	$=N-$ (ring)	6.03	4.71
$>N-$	0.91	0.41	$-S-$ (ring)	−0.6	2.6
$=N-$	1.93	3.24	$-CO-O-CO-$ (anidride)	10.6	11.49

Example 1.3: Use eqs (1.5) and (1.6) and calculate the FP of the following isomers:
(a) 2-Formylbenzoic acid; NBP = 561 K [43]

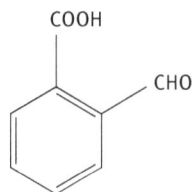

(b) 4-Carboxybenzaldehyde; NBP = 702 K [43]

Answer: According to Table 1.3 and eqs (1.5) and (1.6), these compounds have one –CHO (aldehyde), one –COOH (acid), four =CH– (ring), and two =C< (ring) functional groups, which provide the FP of these compounds as follows:

(a) Equation (1.5):

$$FP = 12.14 + 0.73\,(NBP) + 1\,(-CHO) + 1\,(-COOH) + 4\,(=CH-\,(ring)) + 2\,(=C<(ring))$$

$$= 12.14 + 0.73\,(702) + 1(4.92) + 1(15.33) + 4(-1.4) + 2(-0.39) = 538.47\ K$$

Equation (1.6):

$$FP = 11.07 + 0.72\,(NBP) + 1(-CHO) + 1(-COOH) + 4(=CH-\,(ring)) + 2(=C<(ring))$$

$$= 11.07 + 0.72(702) + 1(3.39) + 1(23.4) + 4(-1.11) + 2(-1.13) = 536.6\ K$$

(b) Equation (1.5):

$$FP = 12.14 + 0.73\,(NBP) + 1(-CHO) + 1(-COOH) + 4(=CH-\,(ring)) + 2(=C<(ring))$$

$$= 12.14 + 0.73(561) + 1(4.92) + 1(15.33) + 4(-1.4) + 2(-0.39) = 435.54\ K$$

Equation (1.6):

$$FP = 11.07 + 0.72\,(NBP) + 1(-CHO) + 1(-COOH) + 4(=CH-\,(ring)) + 2(=C<(ring))$$

$$= 11.07 + 0.72(561) + 1(3.39) + 1(23.4) + 4(-1.11) + 2(-1.13) = 435.08\ K$$

The experimental data of the FPs of 4-carboxybenzaldehyde and 2-formylbenzoic acid are 552 and 440 K [43], respectively. For eq. (1.5), the resultant errors for 4-carboxybenzaldehyde and 2-formylbenzoic acid are 12 and 4 K, respectively. For eq. (1.6), the resultant deviations for 4-carboxybenzaldehyde and 2-formylbenzoic acid are 13 and 5 K, respectively.

1.2.1.5 Non-linear model for the prediction of FP using NBP

Serat et al. [48] used a large dataset of 1,660 components from different material classes to find the following equation, which has good extrapolation ability and correlative power:

$$FP = 2.9513 \times NBP \times \left(\frac{0.33156 + \sum_i N_i C_i + \sum_j M_j D_j + \sum_k E_k O_k}{1 + \left|0.33156 + \sum_i N_i C_i + \sum_j M_j D_j + \sum_k E_k O_k\right|} \right) \quad (1.7)$$

where the NBP is in K; N_i, M_j, and E_k are the numbers of occurrences of individual group contributions; C_i is the first-order group contribution of type i; D_j is the second-order group contribution of type j, and O_k is the third-order group contributions of the type k. All parameters except NBP are dimensionless. Tables 1.4–1.6 give the values of C_i, D_j, and O_k.

Example 1.4: Use eq. (1.7) and calculate the FPs of the following organic compounds:
(a) o-Terphenyl; NBP = 609.15 K [43]

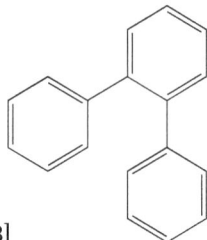

(b) Trilactic acid; NBP = 619 K [43]

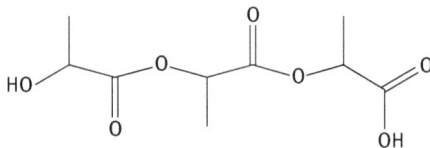

(c) N,N'-Di-tert-butylethylenediamine; NBP = 462.15 K [43]

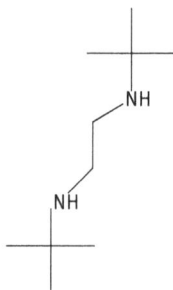

(d) Methylcyclopentadiene Dimer; NBP = 473 K [43]

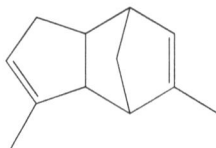

(e) Pentafluoroethyl Trifluorovinyl Ether; NBP = 281.15 K [43]

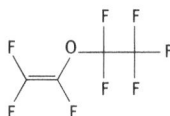

Answer: The use of eq. (1.7) and Tables 1.4–1.6 can predict the FP for each part as follows:

(a)

	First-order group contributions		
Groups	N_i	C_i	$\sum_i N_i C_i$
aCH	14	−0.00050	14 × (−0.00050)
aC except as above	4	0.00290	4 × (0.00290)
	Second-order group contributions		
Groups	M_j	D_j	$\sum_j M_j D_j$
AROMRINGs^{1}s^2	1	−0.00269	1 × (−0.00269)
	Third-order group contributions		
Groups	E_k	O_k	$\sum_k E_k O_k$
aC-aC (different rings)	2	−0.00598	2 × (−0.00598)

$$\sum_i N_i C_i + \sum_j M_j D_j + \sum_k E_k O_k = -0.01008$$

$$FP = 2.9513 \times NBP \times \left(\frac{0.33156 + \sum_i N_i C_i + \sum_j M_j D_j + \sum_k E_k O_k}{1 + \left| 0.33156 + \sum_i N_i C_i + \sum_j M_j D_j + \sum_k E_k O_k \right|} \right)$$

$$= 2.9513 \times 609.15 \times \left(\frac{0.33156 - 0.01008}{1 + |0.33156 - 0.01008|} \right) = 437.36 \text{ K}$$

The measured FP is 436 K [43].

(b)

First-order group contributions			
Groups	N_i	C_i	$\sum_i N_i C_i$
CH_3	3	0.00403	$3 \times (0.00403)$
CH	1	−0.00751	$1 \times (−0.00751)$
OH	1	0.02297	$1 \times (0.02297)$
COOH	1	0.02409	$1 \times (0.02409)$
CHCOO	2	−0.00572	$2 \times (−0.00572)$

Second-order group contributions			
Groups	M_j	D_j	$\sum_j M_j D_j$
CHCOOH or CCOOH	1	0.00524	$1 \times (0.00524)$
$OH-CH_n-COO$ (n in 0..2)	1	0.00133	$1 \times (0.00133)$

Third-order group contributions			
Groups	E_k	O_k	$\sum_k E_k O_k$
–	–	–	–

$$\sum_i N_i C_i + \sum_j M_j D_j + \sum_k E_k O_k = 0.04679$$

$$FP = 2.9513 \times NBP \times \left(\frac{0.33156 + \sum_i N_i C_i + \sum_j M_j D_j + \sum_k E_k O_k}{1 + \left| 0.33156 + \sum_i N_i C_i + \sum_j M_j D_j + \sum_k E_k O_k \right|} \right)$$

$$= 2.9513 \times 619 \times \left(\frac{0.33156 - 0.04679}{1 + |0.33156 - 0.04679|} \right) = 501.47 \text{ K}$$

The measured FP is 499 K [43].

(c)

First-order group contributions			
Groups	N_i	C_i	$\sum_i N_i C_i$
CH_3	6	0.00403	$6 \times (0.00403)$
C	2	−0.02247	$2 \times (−0.02247)$
CH_2NH	2	0.00097	$2 \times (0.00097)$

(continued)

Second-order group contributions			
Groups	M_j	D_j	$\sum\limits_{j} M_j D_j$
$(CH_3)_3C$	2	0.00501	$1 \times (0.00501)$

Third-order group contributions			
Groups	E_k	O_k	$\sum\limits_{k} E_k O_k$
–	–	–	–

$\sum\limits_{i} N_i C_i + \sum\limits_{j} M_j D_j + \sum\limits_{k} E_k O_k = -0.0088$

$$FP = 2.9513 \times NBP \times \left(\frac{0.33156 + \sum\limits_{i} N_i C_i + \sum\limits_{j} M_j D_j + \sum\limits_{k} E_k O_k}{1 + \left| 0.33156 + \sum\limits_{i} N_i C_i + \sum\limits_{j} M_j D_j + \sum\limits_{k} E_k O_k \right|} \right)$$

$$= 2.9513 \times 462.15 \times \left(\frac{0.33156 - 0.0088}{1 + |0.33156 - 0.0088|} \right) = 332.82 \text{ K}$$

The measured FP is 336 K [43].

(d)

First-order group contributions			
Groups	N_i	C_i	$\sum\limits_{i} N_i C_i$
CH_3	2	0.00403	$2 \times (0.00403)$
CH_2 (cyclic)	2	-0.00068	$2 \times (-0.00068)$
CH (cyclic)	4	-0.00150	$4 \times (-0.00150)$
CH=C (cyclic)	2	-0.00334	$2 \times (-0.00334)$

Second-order group contributions			
Groups	M_j	D_j	$\sum\limits_{j} M_j D_j$
$(CH_n=C)_{cyc}-CH_3$ (n in 0..1)	4	-0.00578	$4 \times (-0.00578)$

Third-order group contributions			
Groups	E_k	O_k	$\sum\limits_{k} E_k O_k$
CH $_{multiring}$	4	0.00182	$4 \times (0.00182)$

$\sum\limits_{i} N_i C_i + \sum\limits_{j} M_j D_j + \sum\limits_{k} E_k O_k = -0.01026$

$$FP = 2.9513 \times NBP \times \left(\frac{0.33156 + \sum_i N_i C_i + \sum_j M_j D_j + \sum_k E_k O_k}{1 + \left| 0.33156 + \sum_i N_i C_i + \sum_j M_j D_j + \sum_k E_k O_k \right|} \right)$$

$$= 2.9513 \times 473 \times \left(\frac{0.33156 - 0.01026}{1 + |0.33156 - 0.01026|} \right) = 339.47 \text{ K}$$

The measured FP is 326.15 K [43].

(e)

First-order group contributions			
Groups	N_i	C_i	$\sum_i N_i C_i$
C=C	1	−0.03340	1 × (−0.03340)
C–O	1	−0.02117	1 × (−0.02117)
CF$_3$	1	0.03262	1 × (0.03262)
−F except as above	5	0.02605	5 × (0.02605)
Second-order group contributions			
Groups	M_j	D_j	$\sum_j M_j D_j$
CH$_m$–O–CH$_n$=CH$_p$ (m,n,p in 0..3)	1	0.00065	1 × (0.00065)
Third-order group contributions			
Groups	E_k	O_k	$\sum_k E_k O_k$
aC-aC (different rings)	2	−0.00598	2 × (−0.00598)

$\sum_i N_i C_i + \sum_j M_j D_j + \sum_k E_k O_k = 0.0878$

$$FP = 2.9513 \times NBP \times \left(\frac{0.33156 + \sum_i N_i C_i + \sum_j M_j D_j + \sum_k E_k O_k}{1 + \left| 0.33156 + \sum_i N_i C_i + \sum_j M_j D_j + \sum_k E_k O_k \right|} \right)$$

$$= 2.9513 \times 281.15 \times \left(\frac{0.33156 - 0.0878}{1 + |0.33156 - 0.0878|} \right) = 245.15 \text{ K}$$

The measured FP is 233 K [43].

Table 1.4: List of the first-order groups, their contributions to the FP, and their number of occurrences in the molecules.

Groups	Frequency of occurrence	C_i	Description	Examples
CH_3	1090	0.00403	$-CH_3$ (methyl group) except as below	Propane (2); Isobutane(3)
CH_2	793	-0.00120	$-CH_2-$ except as below	n-Butane (2); n-Hexane(4)
CH	236	-0.00751	$>CH-$ except as below	2,3-Butanediol (2); 3-Ethylpentane(1)
C	81	-0.02247	$>C<$ except as below	Neopentane (1); 2,2,3,3-Tetramethylhexane(2)
$CH_2=CH$	107	0.00226	$CH_2=CH-$ except as below	1-Nonene (1); 1,5-Hexadiene(2)
CH=CH	67	-0.00210	$-CH=CH-$ except as below	trans-2-Eicosene(1); cis,trans-2,4-Hexadiene (2)
$CH_2=C$	46	-0.00593	$CH_2=C<$ except as below	Isobutane(1); 2,5-DimethyL-1,5-Hexadiene(2)
CH=C	18	-0.01332	$-CH=C<$ except as below	2-Methyl-2-Butene (1); 2,5-Dimethyl-2,4-Hexadiene (2)
C=C	8	-0.03340	$>C=C<$ except as below	Hexachloro-1,3-Butadiene (2); Chlorotrifluoroethylene(1)
$CH_2=C=CH$	3	-0.00174	$CH_2=C=CH-$	1,2-Butadiene (1); 1,2-Hexadiene(1)
$CH_2=C=C$	1	-0.00525	$CH_2=C=C<$ except as below	3-Methyl-1,2-Butadiene (1);
$CH\equiv C$	13	0.00078	$CH\equiv C-$ except as below	1-Octyne(1); Propargyl chloride(1)
$C\equiv C$	6	0.00094	$-C\equiv C-$ except as below	1-Pentene-3-yne (1); pent-4-en-2-yn-1-ol (1)
aCH	440	-0.00050	$=CH-$ aromatic carbon	Monomethyl terephthalate (4); 2-(alpha-Methylbenzyloxy)-1-Propanol (5)
aC fused with aromatic ring	35	-0.00631	fused aromatic carbon atom with an aromatic ring	8-Hydroxyquinoline (2); 2,6-Naphthalenedicarboxylic acid(2)
aC fused with non-aromatic ring	44	-0.00089	fused aromatic carbon atom with a non-aromatic ring	4,6-Dimethyldibenzothiophene(4); 2-Mercaptobenzothiazole(2)
aC except as above	42	0.00290	aromatic carbon except as above	Diphenylacetylene(1); m-Terphenyl(4)
aN in aromatic ring	18	0.00650	aromatic nitrogen atom	Pyrazine(2); Melamine(3)
$aC-CH_3$	98	0.00105	aromatic carbon connected to CH_3	Quinaldine(1); 4-Chloro-o-Xylene(2)
$aC-CH_2$	94	-0.00360	aromatic carbon connected to CH_2	Ethylbenzene(1); o-Diethylbenzene(2)

(continued)

Table 1.4: (continued)

Groups	Frequency of occurrence	C_i	Description	Examples
aC–CH	33	−0.00127	aromatic carbon connected to CH	1,3,5-Triisopropylbenzene(3); 1,2,4,5-Tetraisopropylbenzene(4)
aC–C	28	−0.02136	aromatic carbon connected to C	1,4-Di-tert-Butylbenzene(2); 1,3,5-Tri-tert-Butylbenzene (3)
aC–CH=CH$_2$	13	−0.00320	aromatic carbon connected to CH=CH$_2$	m-Divinylbenzene (2); p-Divinylbenzene(2)
aC–CH=CH	7	−0.01140	aromatic carbon connected to CH=CH	cis-1-Propenylbenzene(1); trans-1-Propenylbenzene(1)
aC–C=CH$_2$	5	−0.00587	aromatic carbon connected to C=CH$_2$	p-Isopropenylstyrene(1); p-Isopropenyl phenol(1)
aC–C≡CH	1	0.00075	aromatic carbon connected to C≡CH	Ethynylbenzene(1); 1,2-diethynylbenzene (2)
aC–C≡C	1	−0.01010	aromatic carbon connected to C≡C	Diphenylacetylene(1); (prop-1-yn-1-yl)benzene (1)
OH	175	0.02297	–OH (for aliphatic chains) except as below	1,2-Propylene glycol(2); Neopentyl glycol(2)
aC–OH	48	0.01841	aromatic carbon connected to OH	Bisphenol a(2); o-Cresol(1)
COOH	83	0.02409	carboxyl group (C(=O)OH) except as below	Pimaric acid(1); Isopimaric acid(1)
aC–COOH	19	0.01795	aromatic carbon connected to carboxyl group	o-Chlorobenzoic acid(1); Trimellitic acid(3)
CH$_3$CO	29	0.01497	carbonyl connected to CH$_3$	Acetylacetone(2); Acetone(1)
CH$_2$CO	20	−0.00073	carbonyl connected to CH$_2$	3-Pentanone(1); Chloroacetyl chloride(1)
CHCO	3	0.00229	carbonyl connected to CH	Diisopropyl Ketone(1); Dichloroacetyl chloride(1)
aC-CO	9	0.00619	carbonyl connected to aromatic carbon	Benzophenone(1); Acetophenone(1)
CHO	45	0.01328	aldehyde group except as below	Glutaraldehyde(2); Furfural(1)
aC–CHO	13	0.00720	aldehyde connected to aromatic carbon	Terephthaldehyde(2); Benzaldehyde(1)

CH_3COO	35	0.01246	ester connected to CH_3	Methyl acetate(1); n-Propyl acetate(1)
CH_2COO	41	0.00160	ester connected to CH_2	Ethyl n-Butyrate(1); Propionic anhydride(1)
CHCOO	11	-0.00572	ester connected to CH	n-Propyl isobutyrate(1); Trilactic acid(2)
CCOO	2	-0.01985	ester connected to C	Ethyl trimethyl acetate(1); Hydroxypivalyl hydroxypivalate(1)
HCOO	17	0.01206	formate group	Benzyl formate(1); Methyl formate(1)
aC-COO	35	0.00325	>C=O of ether group connected to an aromatic carbon	Dioctyl phthalate(2); Benzyl benzoate(1)
aC-OOC	2	0.05710	-O- of ether group connected to an aromatic carbon	Acetylsalicylic acid (1); Phenyl acetate(1)
COO except as above	36	0.01116	ester group except as above	n-Propyl methacrylate(1); Dimethyl-1,4-cyclohexane dicarboxylate(2)
CH_3O	41	0.00739	-O- connected to CH_3	Methyl ethyl ether(1); Methylal(2)
CH_2O	71	0.00014	-O- connected to CH_2	Acetal(2); Isopropyl isobutyl ether(1)
CH-O	7	-0.00265	-O- connected to CH	Diisopropyl ether(1); Bis(difluoromethyl)ether(1)
C-O	5	-0.02117	-O- connected to C	Di-t-butyl peroxide(2); Di-tert-butyl ether(1)
aC-O	14	0.00386	-O- connected to C	trans-3,5-Dimethoxystilbene(2); Anethole(1)
CH_2NH_2	39	0.01022	CH_2 connected to $-NH_2$	n-Propylamine(1); Isobutylamine(1)
$CHNH_2$	7	0.00663	CH connected to $-NH_2$	Isopropylamine(1); L-phenylalanine(1)
CNH_2	2	-0.00950	C connected to $-NH_2$	2-Methyl-2-aminobutane(1); tert-Butylamine(1)
CH_3NH	3	0.01821	CH_3 connected to >NH	Dimethylamine(1); N-Methylcyclohexylamine(1)
CH_2NH	17	0.00097	CH_2 connected to >NH	Triethylenetetramine(2); Di-n-Butylamine(1)
CHNH	1	0.00040	CH connected to >NH	Diisopropylamine(1); N-[1-(isopropylamino)ethyl] hydroxylamine (2)
CH_3N	8	-0.00177	CH_3 connected to >N-	Dimethylethanolamine(1); Tetramethylethylenediamine (2)
CH_2N	9	-0.01328	CH_2 connected to >N-	Triethylamine(1); Triallylamine(1)

(continued)

Table 1.4: (continued)

Groups	Frequency of occurrence	C_i	Description	Examples
aC–NH$_2$	24	0.01087	aromatic carbon connected to NH$_2$	Melamine(3); o-Phenylenediamine(2)
aC–NH	9	0.00160	aromatic carbon connected to NH	N,N'-DIphenyl-p-Phenylenediamine(2); p-Nitrodiphenylamine(1)
aC–N	4	−0.00839	aromatic carbon connected to >N–	N,N-Dimethylaniline(1); p-Dimethylaminobenzaldehyde (1)
NH$_2$ except as above	9	0.02780	–NH$_2$ except as above	Cyclopentylamine(1); Dicyandiamide(1)
CH$_2$CN	24	0.00795	>CH$_2$ connected to –C≡N	Succinonitrile(2); Hydracrylonitrile(1)
CHCN	4	−0.01075	>CH– connected to –C≡N	Isobutyronitrile(1); Acrolein cyanohydrin(1)
CCN	1	−0.01586	>C< connected to –C≡N	Acetone cyanohydrin(1); 2,2-dihydroxypropanenitrile(1)
aC–CN	2	−0.00223	aromatic carbon connected to –C≡N	Nicotinonitrile(1); Benzonitrile(1)
CN except as above	15	0.03063	–C≡N except as above	Cyanogen(2); Cyanogen chloride(1)
CH$_2$NCO	2	0.01557	–N=C=O connected to >CH$_2$	n-Butyl isocyanate(1); 1,6-Hexamethylene diisocyanate (2)
aC–NCO	6	0.01119	aromatic carbon connected to –N=C=O	2,6-Toluene diisocyanate(2); 3,4-Dichlorophenyl isocyanate(1)
CH2NO$_2$	3	0.01495	–NO$_2$ (nitro group) connected to >CH$_2$	Nitroethane(1); 1-Nitropropane(1)
CHNO$_2$	1	0.01141	–NO$_2$ (nitro group) connected to >CH–	2-Nitropropane(1); 2-nitropentane (1)
aC–NO$_2$	29	0.00998	–NO$_2$ (nitro group) connected to aromatic carbon	m-Dinitrobenzene(2); o-Nitrodiphenylamine(1)
NO$_2$ except as above	1	0.04834	–NO$_2$ (nitro group) except as above	Nitromethane(1);
CONH$_2$	8	0.03077	–CO–NH$_2$ (amide group)	Hexanamide(1); Trifluoroacetamide(1)
CONHCH$_3$	1	0.03288	–CO–NH– connected to CH$_3$	N-methylacetamide(1);
CON(CH$_3$)$_2$	1	0.01404	–CO–N< connected to two CH$_3$	N,N-Dimethylacetamide(1);

aC−CONH$_2$	1	0.02264	aromatic carbon connected to −CO−NH$_2$ (amide group)	Benzamide(1);
aC−NH(CO)H	1	0.02115	aromatic carbon connected to −NH−C(=O)H	Formanilide(1);
aC−NHCO	3	−0.01224	aromatic carbon connected to −NH−C=O	Acetaminophen(1); Acetanilide(1)
NHCONH	1	0.02663	>C=O connected to two >NH	1,3-Dimethyl urea(1);
NH$_2$CONH	1	0.05138	>C=O connected to both >NH and −NH$_2$	Monomethyl urea(1);
CH$_2$Cl	41	0.01481	−Cl connected to >CH$_2$	1,4-Dichloro-trans-2-Butene(2); Ethyl chloride
CHCl	8	0.00357	−Cl connected to >CH-	2,3-Dichlorobutane(2); 2,3-Dichloro-1-Propanol(1) tert-Butyl chloride(1);
CCl	1	−0.01124	−Cl connected to C	
CHCl$_2$	5	0.03679	−CH connected to two Cl	Dichloroacetaldehyde(1); 1,1-Dichloroethane(1)
CCl$_3$	5	0.05920	−C connected to three Cl	Trichloroacetaldehyde(1); 1,1,1-Trichloroethane(1)
CH$_2$F	6	0.01619	CH$_2$ connected to F	1,2-Difluoroethane(2); 1,1,1-Trichlorofluoroethane
CHF$_2$	6	0.02531	−CH connected to two F	Difluoromethyl methyl ether(1); 1,1-Difluoroethane(1)
CF$_2$	2	0.00761	−C connected to two F	Pentafluoroethyl methyl ether(1); 1,1,2,2,3-Pentafluoropropane(1)
CF$_3$	10	0.03262	−C connected to three F	Pentafluoroethyl trifluorovinyl ether(1); Trifluoroacetamide
CCl$_2$F	1	0.04787	−C connected to both Cl$_2$ and F	1,1-Dichloro-1-fluoroethane(1);
CClF$_2$	1	0.03432	−C connected to both Cl and F$_2$	1-Chloro-1,1-Difluoroethane(1);
aC−Cl	32	0.01161	aromatic carbon connected to Cl	o-Dichlorobenzene(2); 3-Chloro-o-xylene(1)
aC−F	3	0.00561	aromatic carbon connected to F	Pentafluorophenol(5); Fluorobenzene(1)
aC−I	1	0.00764	aromatic carbon connected to I	Iodobenzene(1);

(continued)

Table 1.4: (continued)

Groups	Frequency of occurrence	C_i	Description	Examples
aC–Br	5	0.00730	aromatic carbon connected to Br	m-Dibromobenzene(2); p-Bromotoluene(1)
–I except as above	6	0.02396	iodine except as above	Diiodomethane(2); n-Butyl iodide(2)
–Br except as above	15	0.01665	bromine except as above	1,1,2,2-Tetrabromoethane(4); 1,2-Dibromododecane(2)
–F except as above	18	0.02605	fluorine except as above	p-Chlorobenzotrifluoride(3); 2-Chloro-1,1-Difluoroethylene(2)
–Cl except as above	49	0.01862	chlorine except as above	cis-1,2-Dichloroethylene(2); Methyl chloride(1)
CHNOH	1	0.03838	$>C=N-OH$ aldoxime (oxime)	Acetaldoxime(1);
CNOH	1	0.02472	$-HC=N-OH$ ketoxime (oxime)	2-Butoxime(1);
OCH$_2$CH$_2$OH	27	0.01419	$-O-CH_2-CH_2-OH$	TRIETHYLENE GLYCOL(2); Diethylene glycol(1)
OCHCH$_2$OH	5	0.00830	$-O-CH-CH_2-OH$	Propylene glycol 2-tert-Butyl ether(1); 2-Methoxy propanol-1(1)
OCH$_2$CHOH	19	0.00519	$-O-CH_2-CH-OH$	Tripropylene glycol(2); Dipropylene glycol(1)
–O–OH	13	0.03754	$-O-OH$	Peracetic acid(1); Methyl hydroperoxide(1)
CH$_2$SH	21	-0.00236	$-CH_2-SH$	Methyl-3-Mercaptopropionate(1); 1,2-Ethanedithiol(2)
CHSH	2	-0.00586	$-CH-SH$	sec-Butyl mercaptan(1); Isopropyl mercaptan(1)
CSH	3	-0.01505	$-C-SH$	tert-Nonyl mercaptan(1); tert-Octyl mercaptan(1)
aC–SH	1	0.00698	aromatic carbon connected to –SH	Phenyl mercaptan(1);
–SH except as above	4	0.02324	–SH except as above	Methyl mercaptan(1); Cyclohexyl mercaptan(1)
CH$_3$S	11	0.00430	$-CH_3$ connected to $-S-$ (thioether)	Dimethyl disulfide(2); Methyl ethyl sulfide(1)
CH$_2$S	20	-0.00592	$-CH_2$ connected to $-S-$ (thioether)	Diethyl disulfide(2); Ethylthioethanol(1)
CHS	2	-0.01849	$-CH$ connected to $-S-$ (thioether)	Diisopropyl sulfide(1); Malathion(1)
CS	3	-0.02504	$-C-$ connected to $-S-$ (thioether)	DI-tert-butyl disulfide(2); DI-tert-butyl sulfide(1)

aC–S–	2	0.00667	aromatic carbon connected to –S– (aromatic thioether)	Methyl phenyl sulfide(1); Diphenyl disulfide(2)
SO	1	0.01760	–S(=O)– (sulfoxide)	Dimethyl sulfoxide(1);
SO$_2$	2	-0.00096	–S(=O)$_2$– (sulfone) non-cyclic	DI-n-Propyl sulfone(1); DI-n-Butyl sulfone(1)
aC-P	1	-0.02074	phosphorus connected to aromatic carbon	Triphenylphosphine(1);
C$_2$H$_3$O	7	0.01584	–CH$_2$–CH–O– cyclic ether with a three-atom ring (epoxide)	alpha-Epichlorohydrin(1); Allyl glycidyl ether(1)
C$_2$H$_2$O	1	0.01065	–CH$_2$-C-O- cyclic ether with a three-atom ring (epoxide)	1,2-Epoxy-2-Methylpropane(1);
CH$_2$ (cyclic)	206	-0.00068	–CH$_2$– in a ring	Bicyclohexyl(1); 1-cis-2-trans-3-Trimethylcyclopentane (2)
CH (cyclic)	131	-0.00150	>CH– in a ring	1,2,3,4-Tetramethylcyclohexane(4); 1-cis-2-trans-4-Trimethylcyclopentane(3)
C (cyclic)	31	-0.00404	>C< in a ring	Tetramantane(2); 1,1-Dimethylcyclopentane(1)
CH=CH (cyclic)	47	-0.00122	–CH=CH– in a ring	1,3-Cyclohexadiene(2); Cyclohexene(1)
CH=C (cyclic)	42	-0.00334	–CH=C< in a ring	Methylcyclopentadiene dimer(2); 1-Methylcyclopentene (1)
C=C (cyclic)	6	-0.00199	>C=C< in a ring	Tetrachlorothiophene(2); Palustric acid(1)
CH$_2$=C (cyclic)	6	0.00158	carbon in a ring double bonded to a sidechain carbon	beta-Pinene(1); Isoagatholal(1)
NH (cyclic)	18	0.01256	>NH in a ring	Piperazine(2); Saccharin(1)
N (cyclic)	14	-0.01117	>N– in a ring	Triethylenediamine(2); N-Methyl-2-Pyrrolidone(1)
CH=N (cyclic)	2	0.01224	–CH=N– in a ring	Pyrazole(1); Isoxazole(1)
C=N (cyclic)	1	0.01402	>C=N– in a ring	2-Mercaptobenzothiazole(1);

(continued)

Table 1.4: (continued)

Groups	Frequency of occurrence	c_i	Description	Examples
O (cyclic)	46	0.01169	-O- in a ring	1,4-Dioxane(2); alpha-Tocopherol(1)
CO (cyclic)	26	0.01082	>C=O in a ring	2-Pyrrolidone(1); Anthraquinone(2)
S (cyclic)	32	0.00670	-S- in a ring	2-Mercaptobenzothiazole(1); Thiophene(1)
SO_2 (cyclic)	3	0.02174	-S(=O)2- (sulfone) cyclic	3-Methyl sulfolane(1); Sulfolane(1)
>NH	4	0.01765	>NH except as above	Hexamethyldisilazane(1); Dicyandiamide(1)
-O-	4	-0.04086	-O- except as above	Dicumyl peroxide(2); Divinyl ether(1)
-S-	1	0.46297	-S- except as above	Dicyclohexyl sulfide(1);
>CO	4	0.01025	>C=O except as above	2-Hydroxypropyl acrylate(1); 2-Hydroxyethyl methacrylate(1)
SiHO	1	-0.00787	>SiH- connected to -O-	Trimethoxysilane(1);
SiO	27	-0.00924	>Si< connected to -O-	BIS[3-(triethoxysilyl)propyl]disulfide(2); Dimethyldimethoxysilane(1)
SiH_2	3	0.00643	-SiH₂- primary silicon group	Dichlorosilane(1); Dimethyl silane(1)
SiH	4	-0.00447	>SiH- secondary silicon group	Trichlorosilane(1); Trimethyl silane(1)
Si	24	-0.02604	>Si< tertiary silicon group	Hexamethyldisilazane(2); Hexamethyldisiloxane(1)
C_{cyclic}=N-	1	-0.00825	carbon in a ring double bonded to a sidechain nitrogen	Cyclohexanone oxime(1);
C_{cyclic}=CH-	1	-0.01681	carbon in a ring double bonded to a sidechain =CH-	5-Ethylidene-2-norbornene(1);
C_{cyclic}=C	2	-0.01429	carbon in a ring double bonded to a sidechain =C<	Neoabietic acid(1); Terpinolene(1)
N=N	1	-0.00646	-N=N- in not ring (azo)	p-Aminoazobenzene(1);
C=NH	1	-0.02386	>C=NH in not ring	Dicyandiamide(1); pentan-3-imine (1)
>C=S	1	-0.00468	>C=S in not ring	N-Methylthiopyrrolidone(1);

HCONH	2	0.02699	−NH−C(=O)H not connected to an aromatic atom (amide)	tert-Butylformamide(1); N-Methylformamide(1)
SiH₃	1	0.00735	−SiH₃ silane group	Methyl silane (1); propylsilane(1)
CH=C=CH	1	−0.00248	−CH=C=CH− two cumulated double bonds	2,3-Pentadiene(1); hepta-3,4-diene (1)
OP(=S)O	1	−0.01856	−P(=S)(−O−)−O−	Malathion(1); O,O-dimethyl ethylsulfanylphospho-nothioate (1)

Table 1.5: List of the second-order groups, their contributions to the FP, and their number of occurrences in the molecules.

Groups	Frequency of occurrence	M_j	Description	Examples
$(CH_3)_2CH$	123	-0.00140	CH connected to two methyl group	Isopentane (1); Isobutane(2)
$(CH_3)_3C$	59	0.00501	C connected to three methyl group	Neopentane(2); 2,2-Dimethylbutane(1)
$CH(CH_3)CH(CH_3)$	10	-0.00468	CH–CH present in the longest chain of a hydrocarbon bond with two methyl group neighbors, one on each side	2,3,4-Trimethylpentane(2); 2,3-Dimethylbutane(1)
$CH(CH_3)C(CH_3)_2$	5	-0.00209	CH–C present in the longest chain of a hydrocarbon bond with three methyl group neighbors	2,3,3,4-Tetramethylpentane(2); 2,2,3-Trimethylbutane(1)
$C(CH_3)_2C(CH_3)_2$	4	0.00938	CH–CH present in the longest chain of a hydrocarbon bond with four methyl group neighbors, two on each side	2,2,3,3-Tetramethylpentane(1); 2,3-Dimethyl-2,3-diphenylbutane(1)
$CH_n=CH_m–CH_p=CH_k$ (k,m,n,p in 0..2)	13	-0.00017	Two conjugated double bonds in a chain	Isoprene(1); Chloroprene(1)
$CH_3–CH_m=CH_n$ (m,n in 0..2)	93	-0.00040	–CH$_3$ connected to sp^2 carbon in a chain	cis-2-Butene(2); Propylene(1)
$CH_2–CH_m=CH_n$ (m,n in 0..2)	117	-0.00093	>CH$_2$ connected to sp^2 carbon in a chain	Linoleic acid(4); Triallylamine(3)
$CH_p–CH_m=CH_n$ (m,n in 0..2; p in 0..1)	20	0.00303	>CH– or >C< connected to sp^2 carbon in a chain	Stigmasterol(2); sec-Butenyl glycol ether(1)
CHCHO or CCHO	10	-0.00183	CH or C connected to HC=O (aldehyde group)	2-Methylheptanal(1); 2-Methylpropanal(1)
CH_3COCH_2	18	-0.01645	–CH$_2$–C(=O)–CH$_3$ (ketone connected to both –CH3 and CH$_2$)	Acetylacetone(2); Methyl ethyl ketone(1)
CH_3COCH or CH_3COC	3	-0.01040	ketone connected to both –CH$_3$ and CH or C	3-Methyl-2-pentanone (1); Methyl isopropyl ketone(1)
CHCOOH or CCOOH	13	0.00524	carboxylic acid connected to CH or C	Dichloroacetic acid(1); Trichloroacetic acid(1)

Group		Description	Examples
CH$_3$COOCH or CH$_3$COOC	8 0.00045	CH$_3$–C(=O)O–CH or CH$_3$–C(=O)O–C	Ethylidene diacetate(2); Isopropyl acetate(1)
CO–O–CO	4 0.00286	–O– connected to two ketone	Acetic anhydride(1); Isobutyric anhydride(1)
CHOH	40 0.00047	secondary alcohol	Tripropylene glycol(2); 2-Butanol(1)
COH	9 0.00942	tertiary alcohol	Triacetone alcohol(2); trans-1,8-Terpin(1)
NCCHOH or NCCOH	3 0.01351	N≡C–CH–OH or N≡C–C–OH	Acetone cyanohydrin(1); Lactonitrile(1)
OH–CH$_n$–COO (n in 0...2)	4 0.00133	OH–CH$_2$–C(=O)O– or OH–CH–C(=O)O– or OH–C–C(=O)O–	Methyl lactate(1); Methyl glycolate(1)
CH$_m$(OH)CH$_n$(OH) (m,n in 0...2)	12 -0.00549	carbon-carbon bond with two –OH neighbors, one on each side	Tartaric acid(1); 1,4-Benzenedicarboxylic acid,bis(2,3-dihydroxypropyl)ester(2)
CH$_m$(OH)CH$_n$(NH$_p$) (m,n,p in 0...2)	6 0.00415	carbon-carbon bond with –OH and nitrogen neighbors, one on each side	Monoethanolamine(1); Diethanolamine(2)
CH$_m$(NH$_2$)CH$_n$(NH$_2$) (m,n in 0...2)	2 0.01149	carbon-carbon bond with two –NH$_2$ neighbors, one on each side	Ethylenediamine(1); 1,2-Propanediamine(1)
CH$_m$(NH)CH$_n$(NH$_2$) (m,n in 1..2)	5 -0.00317	carbon-carbon bond with –NH$_2$ and >NH neighbors, one on each side	Tetraethylenepentamine(2); N-Aminoethyl ethanolamine(1)
CH$_m$(NH$_n$)–COOH (m,n in 0...2)	3 0.00875	carbon connected to nitrogen and carboxylic acid	L-Phenylalanine(1); Lysine(1)
HOOC–CH$_n$–COOH (n in 1...2)	1 0.00393	carbon connected to two carboxyl groups	Malonic acid(1); 2-ethylpropanedioic acid (1)
HOOC–CH$_n$–CH$_m$–COOH (n, m in 1..2)	2 -0.03016	carbon-carbon bond with two carboxyl groups, one on each side	Succinic acid(1); Tartaric acid(1)
HO–CH$_n$–COOH (n in 1...2)	5 -0.00421	carbon connected to –OH (hydroxy group) and carboxylic acid	Tartaric acid(2); Hydroxycaproic acid(1)
CH$_3$–O–CH$_n$–COOH (n in 1..2)	1 0.01696	CH$_3$–O– connected to carbon-carboxyl group bond	Methoxyacetic acid(1);

(continued)

Table 1.5: (continued)

Groups	Frequency of occurrence	M_j	Description	Examples
HS–CH–COOH	1	0.02342	HS-carbon bond connected to carboxyl group bond	Thioglycolic acid(1);
HS–CH$_n$–CH$_m$–COOH (n, m in 1..2)	1	-0.02151	HS-carbon bond connected to carbon-carboxyl group bond	3-Mercaptopropionic acid(1);
NC–CH$_n$–CH$_m$–CN (n, m in 1..2)	1	-0.00638	nitrile-carbon bond connected to carbon-nitrile bond … ..	Succinonitrile(1); 2-ethylbutanedinitrile(1)
OH–CH$_n$–CH$_m$–CN (n, m in 1..2)	1	0.01809	HO-carbon bond connected to carbon-nitrile bond	Hydracrylonitrile(1);
HS–CH$_n$–CH$_m$–SH (n, m in 1..2)	1	0.01928	carbon-carbon bond connected to two –SH, one on each side	1,2-Ethanedithiol(1); butane-2,3-dithiol(1)
COO–CH$_n$–CH$_m$–OOC (n, m in 1..2)	4	-0.00315	carbon-carbon bond connected to two carboxyl groups by –O–, one on each side	Glyceryl triacetate(2); 2-Acetoacetoxy ethyl methacrylate(1)
OOC–CH$_n$–CH$_m$–COO (n, m in 1..2)	2	0.00234	carbon-carbon bond connected to two carboxyl groups by >C=O, one on each side	Malathion(1); Diethyl succinate(1)
NC–CH$_n$–COO (n in 1…2)	2	0.01505	carbon connected to nitrile and >C=O of carboxyl group	Methyl cyanoacetate(1); Ethyl cyanoacetate (1)
COCH$_n$COO (n in 1..2)	4	0.01106	carbon connected to >C=O (ketone) and >C=O of ester group	2-Acetoacetoxy ethyl methacrylate(1); t-Butyl acetoacetate(1)
CH$_m$–O–CH$_n$=CH$_p$ (m,n,p in 0..3)	5	0.00065	carbon-oxygen bond connected to carbon-carbon double bond	Methyl vinyl ether(1); Ethyl vinyl ether(1)
CH$_m$=CH$_n$–F (m,n in 0..2)	8	-0.00706	carbon-carbon double bond connected to –F	Tetrafluoroethylene(4); Chlorotrifluoroethylene(3)
CH$_m$=CH$_n$–Br (m,n in 0..2)	1	0.02000	carbon-carbon double bond connected to –Br	Vinyl bromide(1);
CH$_m$=CH$_n$–Cl (m,n in 0..2)	14	0.00217	carbon-carbon double bond connected to –Cl	cis-1,2-Dichloroethylene(2); Vinyl chloride(1)

Group			Description	Examples
$CH_m=CH_n–CN$ (m,n in 0..2)	8	-0.01372	carbon-carbon double bond connected to –C≡N	Fumaronitrile(2); Acrylonitrile(1)
$CH_n=CH_m–COO–CH_p$ (m,n,p in 0..3)	30	-0.00588	carbon-carbon double bond connected to –C(=O)–O–$CH_{0,1or\,2}$	Dibutyl maleate(2); Methyl acrylate(1)
$CH_m=CH_n–CHO$ (m,n in 0..2)	6	0.00244	carbon-carbon double bond connected to aldehyde group	Methacrolein(1); Acrolein(1)
$CH_m=CH_n–COOH$ (m,n in 0..2)	6	0.01061	carbon-carbon double bond connected to carboxyl group	Itaconic acid(1); Acrylic acid(1)
$aC–CH_n–X$ (n in 1..2) X: Halogen	3	-0.00971	aromatic carbon-carbon bond connected to halogen elements	Benzotrichloride(3); Benzyl dichloride(2)
$aC–CH_n–NH_m$ (n in 1..2; m in 0..2)	1	-0.01901	aromatic carbon-carbon bond connected to nitrogen	Benzylamine(1); 1-[2-(aminomethyl)phenyl]methanamine (2)
$aC–CH_n–O–$ (n in 1..2)	10	0.00128	aromatic carbon-carbon bond connected to –O–	BIS(alpha-Methylbenzyl) Ether(2); Benzyl ethyl ether(1)
$aC–CH_n–OH$ (n in 1..2)	8	-0.00019	aromatic carbon-carbon bond connected to –OH	m-Tolualcohol(1); Benzyl alcohol(1)
$aC–CH_n–CN$ (n in 1..2)	1	-0.02201	aromatic carbon-carbon bond connected to –C≡N	Phenylacetonitrile(1);
$aC–CH_n–CHO$ (n in 1..2)	1	-0.01711	aromatic carbon-carbon bond connected to aldehyde	2-Phenylpropionaldehyde(1);
$aC–CH_n–SH$ (n in 1..2)	1	-0.02142	aromatic carbon-carbon bond connected to –SH	Benzyl mercaptan(1); [4-(sulfanylmethyl)phenyl]methanethiol(2)
$aC–CH_n–COOH$ (n in 1..2)	2	-0.00343	aromatic carbon-carbon bond connected to –COOH	4-Methoxyphenylacetic acid(1); Ibuprofen(1)
$aC–CH_n–OOC–H$ (n in 1..2)	2	-0.00489	aromatic carbon-carbon bond connected to –O–C(H)=O	alpha-Methylbenzyl alcohol formate(1); Benzyl formate(1)

(continued)

Table 1.5: (continued)

Groups	Frequency of occurrence	M_j	Description	Examples
aC-CH$_n$-OOC-H (n in 1..2)	2	-0.00105	aromatic carbon-carbon bond connected to -O-C=O	Benzyl benzoate(1); Benzyl acetate(1)
aC-CH$_n$-COO (n in 1..2)	1	-0.00113	aromatic carbon-carbon bond connected to -C(=O)-O-	Ethyl phenyl acetate(1);
aC-CH(CH$_3$)$_2$	16	-0.00859	aromatic carbon-carbon bond connected to two -CH$_3$	1,3,5-Triisopropylbenzene(3); 1,2,4,5-Tetraisopropylbenzene(4)
aC-C(CH$_3$)$_3$	8	0.00461	aromatic carbon-carbon bond connected to three -CH$_3$	1,4-DI-tert-Butylbenzene(2); 1,3,5-TRI-tert-Butylbenzene(3)
aC-CF$_3$	5	-0.04778	aromatic carbon-carbon bond connected to three -F	Benzotrifluoride(1); 3-Nitrobenzotrifluoride(1)
(CH$_n$=C)($_{cyc}$)-CHO (n in 0..1)	1	-0.00120	-CH=C< or >C=C< in a ring connected to -C(H)=O	Furfural(1);
(CH$_n$=C)$_{cyc}$-CH$_3$ (n in 0..1)	27	-0.00578	-CH=C< or >C=C< in a ring connected to -CH$_3$	Methylcyclopentadiene dimer(2); 2,3-Dimethylthiophene(2)
(CH$_n$=C)$_{cyc}$-CH$_2$ (n in 0..1)	9	-0.00309	-CH=C< or >C=C< in a ring connected to -CH$_2$	Furfuryl alcohol(1); Propenyl cyclohexene(1)
(CH$_n$=C)$_{cyc}$-Cl (n in 0..1)	1	-0.01165	-CH=C< or >C=C< in a ring connected to -Cl	Tetrachlorothiophene(4);
CH$_{cyc}$-CH$_3$	44	-0.00451	>CH- in a ring connected to -CH$_3$	Methylcyclopentane(1); cis-1,2-Dimethylcyclopentane(2)
CH$_{cyc}$-CH$_2$	25	-0.00448	>CH- in a ring connected to non ring -CH$_2$-	trans-1,4-Diethylcyclohexane(2); Ethylcyclopentane(1)
CH$_{cyc}$-CH	9	-0.00003	>CH- in a ring connected to non ring >CH-	Isopropylcyclopentane(1); Sitosterol(1)
CH$_{cyc}$-C	5	-0.00175	>CH- in a ring connected to non ring >C<	tert-Butylcyclohexane(1); alpha-Terpineol(1)
CH$_{cyc}$-CH=CH$_n$ (n in 1..2)	3	-0.00573	>CH- in a ring connected to non ring -CH =CH$_{1or2}$	Vinylnorbornene(1); Vinylcyclohexene(1)

Group	n	Value	Description	Examples
CH_{cyc}–C=CH_n (n in 1...2)	2	0.00059	>CH– in a ring connected to non ring >C=$CH_{1or\ 2}$	d-Limonene(1); beta-Terpineol(1)
CH_{cyc}–OH	12	-0.00677	>CH– in a ring connected to –OH	Inositol(6); beta-Cholesterol(1)
CH_{cyc}–NH_2	3	-0.01616	>CH– in a ring connected to –NH_2	Cyclopentylamine(1); Cyclopropylamine(1)
CH_{cyc}–NH–CH_n (n in 0..3)	2	-0.01302	>CH– in a ring connected to non ring NH– carbon bond	Dicyclohexylamine(2); N-Methylcyclohexylamine(1)
CH_{cyc}–SH	1	-0.02513	>CH– in a ring connected to –SH	Cyclohexyl mercaptan(1);
CH_{cyc}–CN	1	-0.01969	>CH– in a ring connected to –C≡N	Cyclopropyl cyanide(1);
CH_{cyc}–COOH	3	0.00135	>CH– in a ring connected to –COOH	1,4-Cyclohexanedicarboxylic acid(2); Cyclopropane carboxylic acid(1)
CH_{cyc}–CO	1	0.00571	>CH– in a ring connected to –C=O	Cyclopropanecarboxamide(1);
CH_{cyc}–S–	1	-0.23428	>CH– in a ring connected to –S–	Dicyclohexyl sulfide(2);
CH_{cyc}–CHO	2	-0.01054	>CH– in a ring connected to –C(H)=O	1,2,3,6-Tetrahydrobenzaldehyde(1); Cyclohexanecarboxaldehyde(1)
CH_{cyc}–O–	3	-0.00490	>CH– in a ring connected to –O–	Cyclohexyl hydroperoxide(1); Methoxydihydropyran(1)
CH_{cyc}–OOCH	1	-0.00162	>CH– in a ring connected to –O–C(H)=O	Cyclohexyl formate(1);
CH_{cyc}–COO	1	-0.00918	>CH– in a ring connected to –C(=O)–O–	Dimethyl-1,4-Cyclohexanedicarboxylate(2);
CH_{cyc}–OOC	2	-0.00457	>CH– in a ring connected to –O–C(H)=O	Acetomethoxane(1); Cyclohexyl acetate(1)
C_{cyc}–CH_3	25	-0.00679	>C< in a ring connected to –CH_3	Isophorone diisocyanate(3); Camphor(2)
C_{cyc}–CH_2–	5	-0.00262	>C< in a ring connected to –CH_2–	1-Methyl-1-Ethylcyclopentane(1); 1,1-Diethylcyclohexane(2)
C_{cyc}–OH	4	-0.00278	>C< in a ring connected to –OH	beta-Terpineol(1); 1-Methylcyclohexanol(1)
>N_{cyc}–CH_3	4	0.01183	>N– in a ring connected to –CH_3	N-Methylpyrrolidine(1); N-Methylpyrrole(1)
>N_{cyc}–CH_2–	8	0.00280	>N– in a ring connected to –CH_2–	4-(2-Aminoethyl)Morpholine(1); N-Ethylmorpholine(1)

(continued)

Table 1.5: (continued)

Groups	Frequency of occurrence	M_j	Description	Examples
AROMRINGs^{1}s^2	58	-0.00269	ortho substitution in benzene	o-Terphenyl(1); o-Cymene(1)
AROMRINGs^{1}s^3	33	-0.00494	meta substitution in benzene	m-Ethyltoluene(1); m-Cresol(1)
AROMRINGs^{1}s^4	73	0.00014	para substitution in benzene	1-(4-Ethylphenyl)-2-(4-Ethylphenyl)Ethane (2); p-Xylene(1)
AROMRINGs^{1}s^{2}s^3	13	-0.00650	substitution in positions 1-2-3 (for benzene)	1,2,3-Trichlorobenzene(1); 2,3-Xylenol(1)
AROMRINGs^{1}s^{2}s^4	33	-0.01117	substitution in positions 1-2-4 (for benzene)	Trioctyl trimellitate(1); 4-Chloro-o-Xylene(1)
AROMRINGs^{1}s^{3}s^5	11	-0.00532	substitution in positions 1-3-5 (for benzene)	1,3,5-Triisopropylbenzene(1); Mesitylene(1)
AROMRINGs^{1}s^{2}s^{3}s^4	3	-0.01075	substitution in positions 1-2-3-4 (for benzene)	1,2,4-Trimethyl-3-Ethylbenzene(1); 1,2,3-Trimethyl-4-Ethylbenzene(1)
AROMRINGs^{1}s^{2}s^{3}s^5	4	-0.01713	substitution in positions 1-2-3-5 (for benzene)	4,6-Dinitro-o-sec-Butylphenol(1); 1,2,3,5-Tetramethylbenzene(1)
AROMRINGs^{1}s^{2}s^{4}s^5	4	-0.00700	substitution in positions 1-2-4-5 (for benzene)	1,2,4,5-Tetramethylbenzene(1); Pyromellitic acid(1)
PYRIDINEs^2	1	-0.00042	substitution in position 2 (for pyridine)	2-Methylpyridine(1); 2-[(pyridin-2-yl)methyl]pyridine (2)
PYRIDINEs^3	3	0.01991	substitution in position 3 (for pyridine)	Niacin(1); Nicotinonitrile(1)
PYRIDINEs^4	1	0.00116	substitution in position 4 (for pyridine)	4-Methylpyridine(1); 4-[(pyridin-4-yl)methyl]pyridine (2)
PYRIDINEs^{2}s^6	1	-0.00766	substitution in positions 2-6 (for pyridine)	2,6-Dimethylpyridine(1);
AROMRINGs^{1}s^{2}s^{3}s^{4}s^5	2	-0.01108	substitution in positions 1-2-3-4-5 (for benzene)	Pentaethylbenzene; Pentamethylbenzene (1)

Table 1.6: List of the third-order groups, their contributions to the FP, and their number of occurrences in the molecules.

Groups	Frequency of occurrence	O_k	Description	Examples
HOOC–(CH$_n$)$_m$–COOH (m>2, n in 0...2)	5	−0.02955	two carboxyl groups connected to each other through linear alkane chains with more than two C	Adipic acid (1); Azelaic acid(1)
NH$_2$–(CH$_n$)$_m$–OH (m>2, n in 0...2)	1	0.01590	−NH$_2$ group connected to −OH group through linear alkane chains with more than two C	3-Amino-1-propanol (1)
OH–(CH$_n$)$_m$–OH (m>2, n in 0...2)	5	−0.00607	2 −OH groups connected to each other through linear alkane chains with more than two C	1,3-Propylene glycol(1); 1,4-Butanediol(1)
NH2–(CH$_n$)$_m$–NH$_2$ (m>2; n in 0...2)	3	−0.00342	2 −NH$_2$ groups connected to each other through linear alkane chains with more than two C	Hexamethylenediamine(1); 1,3-Propanediamine(1)
NC–(CH$_n$)$_m$–CN (m>2)	2	−0.00005	2 cyano-groups connected to each other through linear alkane chains with more than two C	Glutaronitrile(1); Adiponitrile(1)
aC–(CH$_n$=CH$_m$)cyc (fused rings) (n,m in 0,,1)	17	−0.00021	aromatic carbon connected to non-aromatic CH$_n$=CH$_m$ (alkenyl group) in a same ring	Acenaphthalene(2); Indene(1)
aC–aC (different rings)	14	−0.00598	2 aromatic carbons connected outside the rings	Biphenyl(1); p-Terphenyl(2)
aC–CH$_{ncyc}$ (different rings) (n in 0..1)	2	−0.00076	aromatic carbon connected to a carbon in a different ring	1-Phenylindene(1); Cyclohexylbenzene(1)
aC–CH$_{ncyc}$ (fused rings) (n in 0...1)	20	−0.00386	aromatic carbon connected to a non-aromatic carbon in a same ring	1,2,3,4-Tetrahydronaphthalene(2); alpha-Tocopherol(1)

(continued)

Table 1.6: (continued)

Groups	Frequency of occurrence	O_k	Description	Examples
aC–(CH$_n$)$_m$–aC (different rings) (m>1; n in 0..2)	3	0.00052	2 aromatic carbons in different rings connected to each other through linear alkane chains with more than two C	4-(2-phenylethyl)pyridine (1); 1,2-Diphenylethane (1)
CH$_{cyc}$–CH$_{cyc}$ (different rings)	3	−0.00943	two carbons in different rings, connected outside the rings	2-Cyclohexyl cyclohexanone(1); Bicyclohexyl(1)
CH $_{multiring}$	33	0.00182	>CH– in multi rings (>CH– belongs to more than one ring)	Stigmasterol (3); 1,3-Dimethyladamantane(2)
C $_{multiring}$	16	−0.00541	>C< in multi rings (>C< belongs to more than one ring)	Stigmasterol (2); Isoagatholal(1)
aC–CH$_m$–aC (different rings) (m in 0..2)	11	0.00265	2 aromatic carbons in different rings connected to each other through non ring carbon	Triphenylmethane (3); Tetraphenylmethane(6)
aC–((CH$_m$=CH$_n$)–aC (different rings) (m,n in 0..2)	5	0.00400	2 aromatic carbons in different rings connected to each other through non ring carbon-carbon double bond	Triphenylethylene (2); Tetraphenylethylene(4)
aC–CO–aC (different rings)	1	0.07690	2 aromatic carbons in different rings connected to each other through keton (>C=O)	Benzophenone (1)
aC–CH$_m$–CO–aC (different rings) (m in 0..2)	1	−0.08448	2 aromatic carbons in different rings connected to each other through keton-carbon bond	2-hydroxy-1,2-diphenylethanone (1);
aC–CO$_{cyc}$ (fused rings)	5	−0.00881	aromatic carbon connected to a non-aromatic >C=O in a same ring	Anthraquinone(4); Phthalic anhydride(2)
aC–S$_{cyc}$ (fused rings)	16	−0.00559	aromatic carbon connected to a non-aromatic –S– in a same ring	4,6-Dimethyldibenzothiophene(2); Benzothiophene(1)

Group	Value	Description	Examples
aC−NH$_{ncyc}$ (fused rings) (n in 0..1)	3	aromatic carbon connected to a non-aromatic -NH- (or >N-) in a same ring	Phenothiazine(2); Indole(1);
aC−NH−aC (different rings)	6	2 aromatic carbons in different rings connected to each other through nitrogen	N,N'-Diphenyl-p-Phenylenediamine (2); Diphenylamine(1)
aC−(N=CH$_n$)$_{cyc}$ (fused rings) (n in 0..1)	1	aromatic carbon connected to a non-aromatic −N=CH- (or −N=C<) in a same ring.	2-Mercaptobenzothiazole(1);
aC−O−aC (different rings)	1	2 aromatic carbons in different rings connected to each other through oxygen	Diphenyl ether(1);
aC−CHn−O−CHm−aC (different rings) (n,m in 0..2)	1	2 aromatic carbons in different rings connected to each other through carbon-oxygen-carbon bonds	Dibenzyl ether(1);
aC−O$_{cyc}$ (fused rings)	4	aromatic carbon connected to a non-aromatic oxygen in a same ring.	Dibenzofuran(2); 2-Methylbenzofuran(1)
AROM,FUSED[2]	58	benzene ring containing successively 4 aromatic carbon atoms (aCH) followed by two fused aromatic carbon (faC). (aCH−aCH−aCH−faC−faC)	Tetralin (1); Naphthalene(2)
AROM,FUSED[2]s[1]	18	benzene ring containing successively 3 aromatic carbon atoms (aCH) followed by one aC and two fused aromatic carbon (faC). (aCH−aCH−aCH−aC−faC−faC)	5-methyl-1,2,3,4-tetrahydronaphthalene (1); 1,8-dimethylnaphthalene (2)

(continued)

Table 1.6: (continued)

Groups	Frequency of occurrence	O_k	Description	Examples
AROM,FUSED[2]s²	12	0.00464	benzene ring containing successively 2 aromatic carbon atoms (aCH) followed by 1 aC, 1 aCH and 2 fused aromatic carbon (faC). (aCH—aCH—aC—aCH—faC—faC)	6-methyl-1,2,3,4-tetrahydronaphthalene (1); 2,6-Diethylnaphthalene (2)
AROM,FUSED[3]	3	0.00682	benzene ring containing successively 3 aromatic carbon atoms (aCH) followed by 3 fused aromatic carbon (faC). (aCH—aCH—aCH—faC—faC—faC)	1,2,2a,3,4,5-hexahydroacenaphthylene (1); Acenaphthalene(2)
AROM,FUSED[4a]	3	0.02485	benzene ring containing successively 2 fused aromatic carbon (faC), 1 aromatic carbon atom (aCH), 2 faC and 1 aCH. (faC—faC—aCH—faC—faC—aCH)	Anthracene(1); Naphthacene(2)
AROM,FUSED[4p]	4	0.01231	benzene ring containing successively 4 fused aromatic carbon (faC) followed by 2 aromatic carbon atoms (aCH) (faC—faC—faC—faC—aCH—aCH)	Phenanthrene(1); Chrysene(2)
PYRIDINE.FUSED[2]	3	0.00214	pyridine ring containing successively 1 aromatic nitrogen atom (aN), 3 aromatic carbon atoms (aCH) and 2 fused aromatic carbon (faC) (aN—aCH—aCH—aCH—faC—faC)	Quinoline(1); 8-Hydroxyquinoline(1)

Group		Value	Description	Examples
PYRIDINE,FUSED[2-iso]	1	0.00562	pyridine ring containing successively 1 aromatic carbon atom (aCH), 1 aromatic nitrogen atom (aN), 2 aCH and 2 fused aromatic carbon (faC) (aCH—aN—aCH—aCH—faC—faC)	Isoquinoline(1); 5,6,7,8-tetrahydroisoquinoline (1)
PYRIDINE,FUSED[4]	1	0.00486	pyridine ring containing successively 2 fused aromatic carbon (faC), 1 aromatic nitrogen atom (aN), 2 faC and 1 aromatic carbon atom (aCH). (faC—faC—aN—faC—faC—aCH)	Acridine (1); 5,7-diazapentacene (2)
N multiring	1	0.02078	>N— in multi rings (>N— belongs to more than one ring)	Triethylenediamine (2); octahydro-1H-quinolizine (1)
>N—	1	0.00840	>N— except as above	2,2',2"-Nitrilotris-acetonitrile (1); tri-benzylamine (1)
N=C=O	2	0.03915	—N=C=O (isocyanate) connected to non ring or non aromatic atom	Methyl isocyanate (1); Isophorone diisocyanate(1)
Ccyc—N=C=O	1	−0.02272	—N=C=O (isocyanate) connected to ring atom	Cyclohexyl isocyanate(1); Isophorone diisocyanate(1)

1.2.2 The SGC methods

The second category for predicting the FP contains models, which are developed based on the SGC methods. According to the SGC methods, properties are calculated as a function of the number and the type of predefined functional groups constituting the compound. Linear SGC method is the simplest form, which offers the following linear equation to predict the property:

$$\psi = c + \sum n_i \psi_i \qquad (1.8)$$

where n_i and ψ_i are the number and contribution of functional group i, respectively, and c is a constant. The SGC methods [42, 49] are widely used for the prediction of different properties, e.g., solid phase heats of formation [50], gas phase heats of formation [51], heats of fusion and fusion temperature [52, 53], and heats of sublimation [54]. For predicting the FP for some classes of organic compounds, the SGC methods are widely used. Application of the SGC methods for calculation of the FP property of pure compounds was done on the basis of different approaches.

Albahri [55] presented a structural SGC method for predicting the flammability characteristics of pure hydrocarbon fluids that have a significant contribution to the overall flammability characteristics and arrive at the sets of groups. The investigated flammability characteristics include the FP, the autoignition temperature (AIT), and the upper and lower flammability limits of about 500 different substances. The calculated values of the SGC method can predict the said flammability properties of pure components from the knowledge of only the molecular structure. The proposed method can predict FP with average percentage error (APE) of 1.8%. Pan et al. [56] constructed relationships between structure and the FP of 92 alkanes by means of an artificial neural network (ANN) using the SGC method. For these alkanes, the average absolute deviation of the predicted FP is 4.8 K, and the root mean square (rms) error being 6.86. Gharagheizi et al. [57] proposed a collection of 79 functional groups to correlate the FP of pure components. This approach constructs a neural network –group contribution correlation to estimate FP of pure components. It was used for predicting the FP of 1,378 pure components of various chemical families. The SGC method lacks accuracy for large datasets of compounds from diverse families. Another drawback of the SGC method is its inability to distinguish between the properties of isomers. Although the SGC methods result in good calculation/prediction of the FP of some classes of organic compounds, their applications are generally limited to a particular group of materials where their values of groups are specified. Some of the best available SGC methods are reviewed here.

1.2.2.1 Organosilicon compounds

Wang et al. [58] proposed a predictive model of the FP for organosilicon compounds via the SGC method. They built up their method by a training set of 184 organosilicon compounds with the average error of 8.91 K. The predictive capability of the proposed model has been demonstrated on a testing set of 46 organosilicon compounds with the average error of 11.15 K. The proposed equation to predict the FPs for organosilicon compounds has the following form:

$$FP = 224.54 + \left(\sum_i v_i f_i \right) - 8.5904 \times 10^{-4} \left(\sum_i v_i f_i \right)^2 \tag{1.9}$$

where v_i is the number of group i in a molecule and f_i is the group contribution for the ith contributed group. The values of f_i are given in Table 1.7.

Example 1.5: Calculate the FP of allyldimethylchlorosilane with the following molecular structure:

Answer: (a) The molecular structure for allyldimethylchlorosilane consists of two $(-CH_3)$, one $(=CH_2)$, one $(=CH-)$, one $(>CH_2)$, one $(>Si<)$, and one $(-Cl$ (attached to Si)).

Table 1.7: The structural group contribution values for organosilicon compounds [58].

Serial No.	Group	f_i	Serial No.	Group	f_i
1	$-CH_3$	−4.7543	21	$-NH_2$	26.7186
2	$>CH_2$	15.6399	22	$>NH$ (non-ring)	17.0723
3	$>CH-$	29.7614	23	$>NH$ (ring)	20.6411
4	$>C<$	41.8024	24	$>N-$ (non-ring)	23.0458
5	$=CH_2$	−0.9360	25	$-N=$ (ring)	58.1676
6	$=CH-$	11.8316	26	$-CN$	46.3254
7	$=C<$	28.2627	27	$-S-$	38.2233
8	$\equiv CH$	25.8011	28	$-SiH_3$	5.1962
9	$\equiv C-$	10.14026	29	$>SiH_2$	7.0559
10	$>CH_2$ (ring)	12.0741	30	$>SiH-$	22.9298
11	$>CH-$ (ring)	26.3594	31	$>Si<$	35.2149
12	$=CH-$ (ring)	11.6540	32	$>SiH-$ (ring)	8.7554
13	$=C<$ (ring)	28.2310	33	$>Si<$ (ring)	21.0877
14	$-F$	−17.3884	34	$>N-$ (ring)	−19.7121
15	$-Cl$	21.5543	35	$-Cl$ (attached to Si)	9.4766
16	$-Br$	29.2475	36	$-I$ (attached to Si)	46.5773
17	$-OH$	49.4208	37	$-N=C=S$	69.7229
18	$-O-$ (non ring)	8.7550	38	$-N=C=O$	18.3904
19	$-O-$ (ring)	17.4888	39	$-N=C=N-$	57.1338
20	$>C=O$ (non-ring) ketone	36.9982			

$$\sum_i v_i f_i = 2(-4.7543) + 1(-0.9360) + 1(11.8316) + 1(15.6399)$$
$$+ 1(35.2149) + 1(9.4766) = 61.7184$$

$$FP = 224.54 + \left(\sum_i v_i f_i\right) - 8.5904 \times 10^{-4} \left(\sum_i v_i f_i\right)^2$$
$$= 224.54 + (61.7184) - 8.5904 \times 10^{-4}(61.7184)^2 = 282.99 \text{ K}$$

The experimental value is 278.15 K [28].

1.2.2.2 MNLR and ANN structural group contribution methods

Albahri [45] used a structural group contribution method to determine the FP using two techniques: multivariable non-linear regression (MNLR) and ANN. The set of 37 atom-type structural groups was used to represent the FP for about 375 substances. The final FP equation has the following form:

$$FP = 180.594 + 23.3514 \left(\sum_i n_i FP_i\right) \tag{1.10}$$

where FP_i is the atom-type structural group contribution and n_i is the number of structural groups in the molecule. The values of FP_i are given in Table 1.8.

Table 1.8: The structural group contribution values [45].

Serial No.	Group	$(FP)_i$	Serial No.	Group	$(FP)_i$
1	$-CH_3$	0.2823	20	$-HC=O$ (aldehyde)	2.0626
2	$>CH_2$	0.6199	21	$-COOH$ (acid)	5.1405
3	$>CH-$	0.8512	22	$=O$	1.8826
4	$>C<$	1.0179	23	$-NH_2$	1.7216
5	$=CH_2$	0.1431	24	$>NH$ (non-ring)	4.7711
6	$=CH-$	0.5856	25	$>N-$ (non-ring)	0.4800
7	$=C<$	0.9265	26	$-O-$ (ring)	0.9161
8	$=C=$	0.6446	27	$>C=O$ (ring)	4.0227
9	$\equiv CH$	0.3692	28	$>NH$ (ring)	2.2231
10	$\equiv C-$	0.7478	29	$-N=$ (ring)	3.4118
11	$>CH_2$ (ring)	0.6125	30	$>N-$ (ring)	1.8971
12	$>CH-$ (ring)	0.6517	31	$-H$	0.0000
13	$>C<$ (ring)	0.4666	32	$>S$	1.8503
14	$=CH-$ (ring)	0.5859	33	$>SO$	7.1659
15	$=C<$ (ring)	1.0747	34	$-SH$	1.2870
16	$-Cl$	0.9295	35	$=S$	2.0413
17	$-OH$	3.2230	36	$\equiv N$	2.8685
18	$-O-$ (non ring)	0.1692	37	$=S=$	0.1067
19	$>C=O$ (non-ring) ketone	2.5228			

Example 1.6: Calculate the FP of (a) p-diethyl benzene and (b) methyl diethanolamine with the following molecular structure:

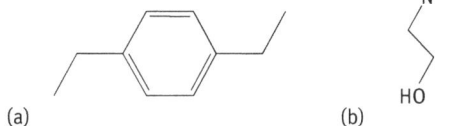

(a)   (b)

Answer: (a) The molecular structure for p-diethyl benzene consists of two (–CH$_3$), two (>CH$_2$), four (=CH (ring)), and two (>C= (ring)).

$$FP = 180.594 + 23.3514 \left(\sum_i n_i FP_i \right)$$
$$= 180.594 + 23.3514 \left(2(0.2823) + 2(0.6199) + 4(0.5859) + 2(1.0747) \right) = 327.7 \text{ K}$$

The experimental value is 329.3 K [59].

(b) The molecular structure for methyl diethanolamine consists one (–CH$_3$), four (>CH$_2$), two (–OH), and one (>N– (non-ring))

$$FP = 180.594 + 23.3514 \left(\sum_i n_i FP_i \right)$$
$$= 180.594 + 23.3514 \left(1(0.2823) + 4(0.6199) + 2(3.2230) + 1(0.4800) \right) = 406.8 \text{ K}$$

The experimental value is 400 K [59].

1.2.3 QSPR models

QSPR methods belong to the third category of methods for predicting the FP in which molecular descriptors, which are numerical quantities calculated from 2-D or 3-D structure of compounds, are used in predictive correlations. Most of QSPR methods are procedures that are more sophisticated. They usually include drawing chemical structure of compounds, minimizing energy, calculating molecular descriptors, and selecting the most effective ones among hundreds of available molecular descriptors. Molecular descriptors are molecular-based parameters where they are numeric characteristics of a pure compound directly calculated from its molecular structure with special algorithms. Several molecular descriptors, normally less than 10 molecular descriptors, are selected to correlate the desired property such as the FP of pure compounds. Several QSPR methods have been introduced in the literature to calculate the FP of pure compounds. Tetteh et al. [60] used radial basis function (RBF) neural network models for the simultaneous estimation of the FP and the NBP based on 25 molecular functional groups and their first-order molecular connectivity index ($^1\chi$) has been developed. The success of this model depends on a network optimization strategy based on biharmonic spline interpolation for the selection of an optimum number of RBF neurons ($\mathbf{n}$) in the hidden layer and their associated spread parameter (σ). The method of Orthogonal Least Squares (OLS) learning algorithm was used for training of the RBF networks. Tetteh et al. [60] divided the total database of 400 compounds into training (134), validation (133), and testing (133), where the average absolute errors (AAEs) obtained for the validation and testing sets are 10 and 12 K, respectively. Katritzky et al. [61] used geometrical, topological, quantum mechanical, and electronic descriptors calculated by CODESSA PRO software to develop QSPR models for predicting the FP of 758 organic compounds. They reported multilinear regression models and a non-linear model based on an artificial neural network. Gharagheizi and Alamdari [62] introduced a general QSPR model for the prediction of the FP of 1,030 pure compounds. They used Genetic Algorithm-based Multivariate Linear Regression (GA-MLR) technique to select four chemical structure-based molecular descriptors from a pool containing 1,664 molecular descriptors. Gharagheizi et al. [34] used experimental NBP of the compound and two chemical structure-based QSPR parameters. They used a comprehensive database of the FPs containing 1,472 pure compounds of various chemical structures for the development of the model. The most important disadvantage of these kinds of QSPR methods is the complex process of calculation for some molecular descriptors from the chemical structure. Molecular descriptors are numerical quantities, which are used in predictive correlations. This kind of QSPR method is less popular than the other methods because of its lower accuracy and more sophisticated procedure. Thus, the selection of appropriate molecular descriptors such as topological indices, quantum chemical parameters, and electrostatic indices, is a key problem. The selected molecular descriptors can be combined with the other methods such as multiple linear regression, partial least squares, and different types of ANNs. These QSPR methods usually require some computer software such as Dragon [63] for obtaining descriptors, which contain unconventional parameters. For all of these QSPR methods, it is important to use a large dataset of different molecular structures because the compounds with similar molecular structure in training set of QSPR procedure should be used as test set. Since the used experimental data for development of the predictive models of the FP are much more than variables, statistical analysis data may be used to confirm the reliability of these methods for those new compounds with similar molecular structures that have not been used in the development of methods. The QSPR correlations based on complex molecular descriptors are not generally simple to develop because these QSPR methods require complex computer codes and expert users.

There are several simple QSPR models based on elemental composition and structural parameters, which have been used to predict the FP of some classes of organic compounds [64–69]. Since most of QSPR methods require complex molecular descriptors and complicated computer codes, the best simple and reliable methods for predicting the FP are demonstrated here where they need only molecular structural parameters.

1.2.3.1 Saturated alkanes

A simple method was introduced for predicting the FP of pure cyclic and acyclic saturated hydrocarbons [69]. It is based on n_C as well as two structural parameters, i.e., increasing (*ISP*) and decreasing (*DSP*) structural parameters. The values of rms and the average absolute deviations of the predicted FPs are 4.6 and 5.4 K for a dataset of 120 and 59 acyclic and cyclic alkanes, respectively. This model is given as:

$$FP = A + 16.15\ n_C + 16.68\ ISP - 24.71\ DSP \qquad (1.11)$$

where A is a constant that is equal to 146.6 and 154.9 for acyclic and cyclic alkanes, respectively. Two correcting functions *ISP* and *DSP* can revise deviations from the predicted results of FPs of saturated hydrocarbons on the basis of n_C. They are specified for saturated cyclic and acyclic hydrocarbons according to the following situations:

ISP – This parameter can be used only for large cycloalkanes that have more than seven-membered rings where it is equal to 1.0 for these compounds.

DSP – This correction parameter may be used for both cyclic and acyclic compounds as:

(i) *Cycloalkanes with three- or four-membered rings:* The value of *DSP* is 1.0 for those compounds without any substituent or with the attachment of only methyl groups.

(ii) *Cycloalkanes with five- or six-membered rings:* For the attachment of large n-alkyl with more than nine carbon atoms $(n_{C>9})$, $DSP = (n_{C>9} - 9) \times 0.3$.

(iii) *Acyclic hydrocarbons with isobutyl molecular fragment (i.e., $(CH_3)_3C-R$):* For alkyl groups containing less than four carbons $(n_{C<4})$, $DSP = 0.3 n_{C<4}$.

(iv) *Small acyclic hydrocarbons:* For $n_C \leq 4$, $DSP = 4.25 - n_{C \leq 4}$.

The parameters *ISP* and *DSP* are equal to zero if the conditions for giving them various values are not met.

Example 1.7: Cyclopentyldodecane has the following molecular structure. Use equation (1.11) and calculate its FP.

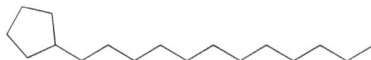

Answer: Since condition (ii) is satisfied, $DSP = (n_{C>9} - 9) \times 0.3 = (12 - 9) \times 0.3 = 0.9$

$$FP = A + 16.15\ n_C + 16.68\ ISP - 24.71\ DSP$$
$$= 154.9 + 16.15 \times 17 + 16.68 \times 0 - 24.71 \times 0.9 = 407.2\ K$$

The measured FP is 409 K, which is taken from the chemical database of the department of chemistry at the University of Akron (USA) [70].

1.2.3.2 Unsaturated hydrocarbons

A simple method was introduced for predicting the FP of different classes of unsaturated hydrocarbons, including alkenes, alkynes, and aromatics. The numbers of carbon and hydrogen atoms are used as a core function that can be revised for some compounds by a correcting function. The rms error is 9 K for a dataset of 173 unsaturated hydrocarbons. The optimized correlation has the following form:

$$FP = 167.1 + (FP)_{core} + (FP)_{correcting} \tag{1.12}$$

where

$$(FP)_{core} = 19.68 n_C - 2.915 n_H \tag{1.13}$$

$$(FP)_{correcting} = 16.77 FP^{(+)} - 32.66 FP^{(-)} \tag{1.14}$$

The quantity n_H is the number of hydrogen atoms. The parameter $(FP)_{core}$ is a linear function of n_C and n_H because it depends on molecular weight and degree of unsaturation of the compound. The factors $FP^{(+)}$ and $FP^{(-)}$ are the contributions of structural parameters of unsaturated hydrocarbons for increasing and decreasing of the FP on the basis of $(FP)_{core}$, respectively, which can be specified according to the following conditions:

(i) $FP^{(+)}$ – This parameter can be applied only to polymethyl benzene where the value of $FP^{(+)}$ depends on the number of methyl groups attached to the benzene ring in the ortho position with respect to each other. Thus, it is equal to $0.25 n_{CH_3}$.

(ii) $FP^{(-)}$ – This correcting function can be applied to aromatic, alkene, and alkyne compounds.

    (a) *Aromatic compounds* – Two different situations can be considered here:

        1. For the attachment of isopropyl directly to the aromatic ring and the presence t-butyl in the molecule, the values of $FP^{(-)}$ are $0.25 n_{isopropyl}$ and $0.5 n_{t-butyl}$, respectively. For example, the value of $FP^{(-)} = 3 \times 0.25 = 0.75$ for 1,3,5-triisopropyl benzene.

        2. For the attachment of large normal alkyl group ($n' \geq 10$), $FP^{(-)}$ equals 1.0 where the parameter n' is the number of carbon atoms in the alkyl group.

    (b) *Alkyne* – For alkynes containing one triple bond with general formula R–C≡C–H, $FP^{(-)}$ equals $1.25 - 0.25 n'$ in which $n' < 5$, where n' is the number of carbon atoms in R substituent, e.g., $FP^{(-)} = 1.25 - 2 \times 0.25 = 0.75$ for 1-butyne.

    (c) *Alkene* – For alkenes with two alkyl groups attached to double bond in form R_1–C=C–R_2, $FP^{(-)}$ is $2.25 - 0.75 n'$ where $n' < 3$. For example, $FP^{(-)}$ is to 1.5 for propene.

    (d) *Two double bonds* – For the existence of two double bonds in form $R_1 - C = C - C = C - R_2$ or $R_1 - C = C = C - R_2$, $FP^{(-)}$ is $1.0 - 0.5 n'$, where only one of the alkyl groups (R_1 or R_2) should be methyl and the other hydrogen atom, e.g., $FP^{(-)} = 0.5$ in 2-methyl butadiene.

Example 1.8: Calculate the FP of tetradecyl benzene with the following molecular structure by using eq. (1.12).

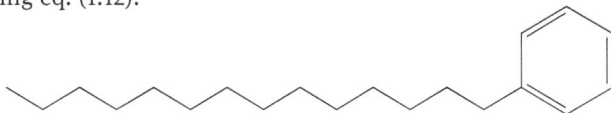

Answer: Since condition 2 of parts (ii) and (a) is satisfied where $n' \geq 10$, $FP^{(-)}=1.0$

$$(FP)_{core} = 19.68n_C - 2.915n_H = 19.68 \times 20 - 2.915 \times 34 = 294.5K$$
$$(FP)_{correcting} = 16.77FP^{(+)} - 32.66FP^{(-)} = 16.77 \times 0 - 32.66 \times 1.0 = -32.66K$$
$$FP = 167.1 + (FP)_{core} + (FP)_{correcting} = 167.1 + 294.5 - 32.66 = 428.9K$$

The reported FP is 433 K [70]. The use of two models Albahri [55] and Rowley et al. [40] give 448 K and 468 K, respectively, which have larger deviations.

1.2.3.3 General correlation between saturated and unsaturated hydrocarbons

A reliable correlation has been introduced for predicting the FP of various types of saturated and unsaturated hydrocarbons containing cyclic and acyclic paraffin, olefins, alkynes, and aromatic hydrocarbons [65]. Large available experimental data consisting of 441 diverse hydrocarbons were used to derive and test the general correlation. For training set containing 423 of these 441 hydrocarbons, the values of rms and average absolute deviations are 7.7 and 5.7 K, respectively. General correlation has the following form:

$$FP = 158.7 + 19.86n_C - 2.40n_H + 51.12P^{(+)} - 49.63P^{(-)} \tag{1.15}$$

where $P^{(+)}$ and $P^{(-)}$ are two correcting functions, which can be easily specified on the basis of some molecular fragments given in the following sections.

1.2.3.3.1 Structural parameter $P^{(+)}$
(i) *Cycloalkanes*: For large cycloalkanes containing more than six-membered rings, the value $P^{(+)}$ is equal to 0.4.
(ii) *Normal alkanes*: For normal alkanes with $n_C \geq 14$, the value of $P^{(+)}$ is 0.25.
(iii) *Methyl-substituted aromatics*: For polymethyl aromatics, the value of $P^{(+)}$ is $0.1n_{CH_3}$ where n_{CH_3} is the number of methyl groups attached to the aromatic ring.

1.2.3.3.2 Structural moieties affecting $P^{(-)}$
(i) *Small cycloalkanes*: For cycloalkanes with three- or four-membered rings without any substituent or with the attachment of only methyl groups, the value of $P^{(-)}$ equals 0.2.

(ii) *Small acyclic hydrocarbons*: If $n_C \leq 5$, then $P^{(-)}=1.40 - (n_C - 1.3) \times 0.35$.

(iii) *The attachment of isopropyl and t-butyl to benzene ring*: The values of $P^{(-)}$ equal 0.85 and 1.5 for the attachment of more than two isopropyl or t-butyl directly to the aromatic ring, respectively.

(iv) *The attachment of normal alkyl group to benzene ring*: The values of $P^{(-)}$ equal 0.5 and 1.0 for the attachment of large normal alkyl group with $10 \leq n' \leq 12$ and $n' > 12$, respectively.

(v) *Alkynes with the general formula* $R-C{\equiv}C-H$: The value of $P^{(-)}$ is $1.16 - 0.18n_C$ where $n_C \leq 6$.

(vi) *Alkenes with formula* $R_1-C=C-R_2$: $P^{(-)}$ equals $1.55 - 0.25n_C$, where $n_C \leq 4$.

(iiv) *The existence of two double bonds*: For the compounds with formula $R_1-C=C-C=C-R_2$ and $R_1-C=C=C-R_2$, the values of $P^{(-)}$ are $1.75 - 0.3n_C$ and $0.95 - 0.1\,n_C$, respectively, for which only one of alkyl groups (R_1 or R_2) should be methyl group and the other hydrogen atom.

1.2.3.3.3 Different behavior of mono-alkyl substituted cyclopentane and cyclohexane

Both $P^{(+)}$ or $P^{(-)}$ can participate in predicting the FP of mono-alkyl substituted cyclopentane and cyclohexane, which depend on the number of carbon atoms in the alkyl group. For the presence of $n' = 1$, 2 and $n' = 3 - 9$, the values of $P^{(+)}$ are 0.1 and 0.2, respectively. Meanwhile, the value of $P^{(-)}$ is $(n'_{>12} - 10) \times 0.05 + 0.05$, where $n'_{>12}$ shows that the number of carbon atoms in the alkyl group is greater than twelve.

The estimated FPs for 18 further hydrocarbons containing complex molecular structures have been compared with the Rowley et al. method [40], which gives much lower values of the rms and average absolute deviations.

Example 1.9: Cyclotetradecane has the following molecular structure. Use eq. (1.15) and calculate its FP.

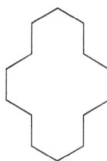

Answer: For this compound, condition (i) in Section 1.2.3.3.1 is satisfied.

$$FP = 158.7 + 19.86n_C - 2.40n_H + 51.12P^{(+)} - 49.63P^{(-)}$$
$$= 158.7 + 19.86 \times 14 - 2.40 \times 28 + 51.12 \times 0.4 - 49.63 \times 0 = 390.0 \text{ K}$$

The measured FP is 433 K [70]. The use of Rowley et al. [40] gives 336 K, which has a larger deviation, i.e., 50 K.

1.2.3.4 Kerosene hydrocarbons

The kerosene fuels are a distillate fraction of petroleum with boiling point between 150 and 300 °C. They are a mixture of hydrocarbons containing compounds with 10 to 16 carbon atoms in both straight chain and branched formations. Zohari and Qhomi [68] presented two new, reliable, simple correlations for predicting FP of kerosene hydrocarbons. Since the reliability of one of them is higher, this correlation is given as:

$$FP = 187.8 + 18.298n_C - 2.570n_H - 4.875n_{R5} + 21.240FP^+ - 26.699FP^- \qquad (1.16)$$

where n_{R5} is the number of five member rings in the molecular formula; FP^+ and FP^- are the several non-additive structural parameters, which are defined as below:

1.2.3.4.1 Definition of FP^+

(1) *Hydrocarbon fuels containing aromatic rings*: For a phenyl ring containing five or six methyl substitutions (e.g., 1,2,3,4,5-pentamethyl benzene), the value of FP^+ is 1.0.
(2) *The compounds containing three or more rings in their structures*: If rings adjoin each other by two or more carbon atoms, two situations are expected:
    (2.1) For six-membered rings and aromatic, $FP^+ = 1.2$.
    (2.2) If there is any ring, which is not a six-membered ring and aromatic, the value of FP^+ is 0.7.

1.2.3.4.2 Description of FP^-:

(1) For polycyclic nonaromatic compounds if there is an unsaturated bond in their structure, $FP^- = 1.0$.
(2) For acyclic hydrocarbon compounds containing two or three quaternary carbon atoms in their structure, the value of FP^- is 0.7.
(3) In the hydrocarbon fuel compounds with aromatic rings, there are several situations:
    (3.1) For the attachment of one or two isopropyl substitutions to the phenyl ring, the value of FP^- is 0.5. Meanwhile, for three isopropyl substitutions, $FP^- = 1.5$.
    (3.2) For the attachment of *tert*-butyl substitution to the phenyl ring, the value of FP^- is 0.7. Meanwhile, if isobutyl or isobutylene are introduced to a phenyl ring, the value of FP^- is 0.5.
    (3.3) For the attachment of more than two ethyl groups to a phenyl ring, the value of FP^- is 0.5.
(4) For the presence of a link as –C–C–, –C=C–, –C≡C– or –C(C)– between two phenyl rings, the value of FP^- is 0.5.

Example 1.10: Find FP of *p*-diisopropyl benzene with the following molecular structure:

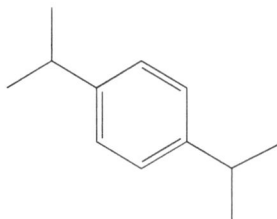

Answer: For this compound, condition (3.1) in Section 1.2.3.4.2 is satisfied.

$$FP = 187.8 + 18.298n_C - 2.570n_H - 4.875n_{R5} + 21.240FP^+ - 26.699$$
$$FP^- = 187.8 + 18.298 \times 12 - 2.570 \times 18 - 4.875 \times 0 + 21.240 \times 0 - 26.699 \times 0.5$$
$$= 347.8K$$

The measured FP is 349 K [43].

1.2.3.5 Various classes of amines

A simple method has been presented for estimating the FP of various types of flammable amines, which include aliphatic amines such as primary, secondary, tertiary, and cyclic amines as well as aromatic amines and heteroarenes containing nitrogen heteroatom. It is based on the contribution of elemental composition and the effects of two correcting functions as:

$$FP = 207.2 + 23.43n_C - 7.363n_H + 49.41n_N + 64.79IP - 62.96DP \tag{1.17}$$

where *IP* and *DP* are increasing and decreasing parameters based on structural parameters of amines. Since FPs of amino derivatives of organic compounds are related to their volatility, the presence of some molecular moieties such as specific polar groups, branching, and the length of substituents attached to amine can affect the values of *IP* and *DP*.

1.2.3.5.1 Prediction of *IP*

(a) *Hydroxyl, chloro, or acyclic ether groups*: For any amino derivative organic compound containing hydroxyl, chloro, or acyclic ether molecular moieties, $IP = n_{OH} + 0.4n_{Cl} + 0.3n_{aO}$ for the presence of hydroxyl (n_{OH}), chloro (n_{Cl}), and acyclic ether (n_{aO}) groups.

(b) *Aromatic compounds containing only $-NH_2$ substituents for which $n_{NH_2} \geq 2$*: The value of *IP* equals 0.4 for these compounds.

(c) *Nitro aniline*: For only nitro-substituted aniline, $IP = 1.0$.

(d) $R-NH_2$ *with $n_C \geq 8$*: The value of *IP* is 0.5 for these alkyl mines.

(e) $-C(=O)-N-$ *or* $-C(=NH)-N-$: For the presence of these groups, $IP = 1.0$.

1.2.3.5.2 Prediction of *DP*

(a) *Primary, secondary, and tertiary amine (only $n_N = 1$) in which each substituent has $n_C \leq 3$*: The value of *DP* is 0.3 for these compounds.

(b) *(Ar)$_2$NH and (Ar)$_3$N*: For aryl amines containing more than one aryl substituent, the values of *DP* are 0.5 and 1.9 for (Ar)$_2$NH and (Ar)$_3$N, respectively.

(c) *Alkyl derivatives of pyridine or pyrrole*: For these compounds, *DP* = 0.6.

Example 1.11: Calculate FP for the following compound:

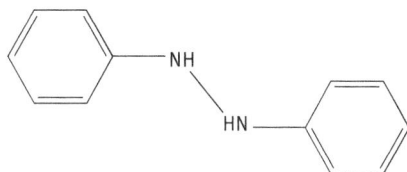

Answer:

$$FP = 207.2 + 23.43n_C - 7.363n_H + 49.41n_N + 64.79IP - 62.96DP$$

$$= 207.2 + 23.43 \times 12 - 7.363 \times 12 + 49.41 \times 2 + 64.79 \times 0 - 62.96 \times 0 = 498.8K$$

The predicted result is close to the measured FP, which is 503 K [43].

1.2.3.6 Alcohols and phenols

A reliable correlation has been developed on the basis of 929 experimental *FP* values of different alcohols and phenols including acyclic and cyclic alcohols as well as phenols and alcohols with composite aliphatic-aromatic structures [67]. It is based on the elemental composition and some of the structural parameters such as intermolecular hydrogen bonding. Among all 929 data points, absolute percent error of this model is greater than 10% only in 24 alcohols and less than 5% in 714 cases. It has the following form:

$$FP = 233.7 + 11.67n_C - 2.028n_H + 18.08n_O + 17.13(n_{Cl} + n_{Br} + n_I) + 33.74HBD$$
$$- 18.56CF \tag{1.18}$$

where n_O, n_{Cl}, n_{Br}, and n_I are the number of oxygen, chlorine, bromine, and iodine atoms, respectively; *HBD* stands for "Hydrogen Bonding Donors" in intermolecular hydrogen bondings that are the number of H atoms attached to O and N atoms, i.e., –OH, –NH$_2$, and >NH functional groups. Final parameter *CF* stands for "Correction Factor" to consider the effects of specific intramolecular attractions, branches, heavier atoms, and long chain alkyl groups on raising or lowering the FPs. The values of *CF* are defined as follows:

(a) *The attachment of –OH, –NH$_2$ or >NH to tertiary carbon*: The value of *CF* equals to the number of hydrogens of these functional groups, which are affected by tertiary carbon steric hindrance. For example, the reported value of the *FP* of 1-butanol is in the range of 302.0–319.3 K but for *tert*-butanol, the *FP* value is between 282.1 and 288.7 K.

(b) *The attachment of large alkyl groups to –OH*: The value of *CF* equals –1.0 for alcohols containing a long chain saturated alkyl group with more than 11 carbon atoms. For unsaturated aliphatic alcohols with more than 11 carbon atoms, the value of *CF* equals +3.

(c) *The presence of C–S–C or –SH groups in aliphatic alcohols*: In these cases, *CF* is equal to –2.0.

Example 1.12: Calculate FP for 2-butyn-1-ol with the following molecular structure:

Answer:

$$FP = 233.7 + 11.67n_C - 2.028n_H + 18.08n_O + 17.13(n_{Cl} + n_{Br} + n_I) + 33.74HBD - 18.56CF$$
$$= 233.7 + 11.67 \times 4 - 2.028 \times 6 + 18.08 \times 1 + 17.13 \times 0 + 33.74 \times 1 - 18.56 \times 0$$
$$= 320.0K$$

The measured value of FP for this compound is 324.8 K [71].

1.2.3.7 Pure organic chemicals from structural contributions

Rowley et al. [40] used experimental data of the FP of more than 1,000 organic compounds to present a suitable method for predicting their FP based solely on structural contributions. The correlation has the following form:

Alcohols or phenols:

$$FP = \frac{\sum_i n_i g_i - 208.30}{2.40 \times \ln(8n_C + 8n_{Si} + 8n_S + 2n_H - 2n_X - 4n_O) + 1} + 196.68 \qquad (1.19)$$

Other Organic compounds:

$$FP = \frac{\sum_i n_i g_i - 510.49}{2.13 \times ln(8n_C + 8n_{Si} + 8n_S + 2n_H - 2n_X - 4n_O) + 1} + 235.21 \qquad (1.20)$$

where n_i is the number of the g_i structural group contribution in the molecule. Equations (1.19) and (1.20) can be used for a wide range of organic compounds but their reliability is lower than previous QSPR models based on elemental composition and structural parameters. Table 1.9 gives structural contributions for eqs (1.19) and (1.20).

Example 1.13: Calculate the FP for 2,2-dimethyl-1-propanol, which was selected from Table 1.10, with the following molecular structure:

Answer: Since this compound is an alcohol, eq. (1.20) should be used as:

Table 1.9: Structural contributions for eqs (1.19) and (1.20).

I	Group	g_i	Example
1	≡C− (HC)	256.43	2-Pentyne
2	≡CH (HC)	261.94	1-Hexyne
3	=C< (HC)	483.40	2-Methyl-1-octene
4	=C_R< (HC)	378.53	d-Limonene
5	=CH− (HC)	219.78	1-Pentene
6	=C_RH− (HC)	124.16	Cyclohexene
7	=CH_2 (HC)	299.53	1-Pentene
8	>C< (HC)	561.32	Neopentane
9	>C_R< (HC)	98.67	1,1-Diethylcyclohexane
10	>CH− (HC)	418.55	4-Methylheptane
11	>C_RH− (HC)	313.87	Methylcyclohexane
12	>CH_2 (HC)	191.61	Pentane
13	−$C_R H_2$− (HC)	122.22	Cyclopentane
14	−CH_3 (HC)	259.62	Propane
15	>CH−	119.79	Isobutyl Formate
16	>C_RH−	201.98	Cyclohexanol
17	>CH_2	162.43	Butanol
18	−$C_R H_2$−	149.72	Cyclohexanol
19	−CH_3	77.80	Ethanol
20	=C<	194.11	Chloroprene
21	=C_R<	236.45	Phenol
22	=C=	2239.01	Carbon Disulfide
23	=CH−	148.59	Acrolein
24	=C_RH−	163.28	Maleic Anhydride
25	=CH_2	37.56	Methacrolein
26	C_R−C_R= (fused ring)	259.24	a-Pinene
27	>C<	108.68	Acetone Cyanohydrin
28	>C_R<	130.20	Isophorone Diisocyanate
29	>C=O	494.20	2-Pentanone
30	>C_R=O	551.77	Cyclopentanone
31	O=CH− (aldehyde)	437.19	Propanal
32	O=$C_R O_R$−	1192.63	Diketene
33	−COO− (ester, nonring)	529.37	Ethyl Acrylate
34	−COOH (acid)	1034.70	Thioglycolic Acid
35	O=$C_R O_R C_R$=O (aliphatic ring)	1750.35	Succinic Anhydride
36	=O	623.68	Di-n-propyl Sulfone

(continued)

Table 1.9: (continued)

I	Group	g_i	Example
37	–O–	176.69	Methyl Ethyl Ether
38	–O$_R$–	128.89	Tetrahydrofuran
39	–OH (alcohol)	803.82	Propanol
40	–OH (phenol)	806.21	Nonylphenol
41	>N–	153.69	Tripropylamine
42	>NH	354.79	Diisopropylamine
43	>N$_R$H	325.82	Ethyleneimine
44	–NH$_2$	362.58	Urea
45	–N=	196.10	Acetaldoxime
46	–N$_R$=	243.93	Oxazole
47	>N$_R$–	369.80	4-(2-Aminoethyl) Morpholine
48	–N–C$_a$	797.35	Diphenylamine
49	–CN	640.67	Ethyl Cyanoacetate
50	–NC=O	1193.67	n-Methylacetamide
51	O=C=N–C$_a$	697.80	Phenyl Isocyanate
52	NO$_2$–C– (aliphatic)	898.17	Nitroethane
53	–NO$_2$	525.91	m-Nitrotoluene
54	–S–	405.65	Diethyl Sulfide
55	–S$_R$–	221.89	Thiophene
56	–SH	469.16	Propyl Mercaptan
57	–Br	386.51	Bromoethane
58	–Cl	251.85	1-Chloropentane
59	–F	255.41	Benzotrifluoride
60	–I	622.38	Hexyl Iodide
61	–Si–	89.55	Dimethyldichlorosilane
62	–O–(Si)	96.01	Hexamethyldisiloxane

HC indicates the group only applies to hydrocarbons, subscript R is an atom belonging to any ring, and C$_a$ is explicitly an aromatic carbon.

Table 1.10: Sample calculations illustrating the estimation of the FP from eqs (1.19) and (1.20).

Chemical	Groups	Group Sum	OH Group	FP$_{calc}$ (K)	FP$_{exp}$ (K)
Terpinolene	3, 4 (×2), 6, 13 (×3), 14 (×3)	1552.42		329.5	329.15
m-Diethylbenzene	4 (×2), 6 (×4), 12 (×2), 14 (×2)	1517.68		327.0	329.0
2,2-Dimethyl-1-propanol	17, 19 (×3), 27, 39	1308.33	39	298.3	303.15
p-Cresol	19, 21 (×2), 24 (×4), 40	2010.03	40	358.6	359.15
Azelaic acid	17 (×7), 34 (×2)	3206.41		491.1	488.15
Dimethylethanolamine	17 (×2), 19 (×2), 39, 41	1437.97	39	315.0	312.15

Table 1.10: (continued)

Chemical	Groups	Group Sum	OH Group	FP$_{calc}$ (K)	FP$_{exp}$ (K)
Tetradecamethylhexa siloxane	19 (×14), 61 (×6), 62 (×5)	2106.55		362.6	375.15
Propylene glycol monoallyl ether	15, 17 (×2), 19, 23, 25, 37, 39	1689.11	39	331.5	327.55
2-Ethyl thiophene	17, 19, 21, 24 (×3), 55	1188.41		302.3	300.15
Maleic anhydride	24 (×2), 30 (×2), 38	1558.99		370.2	375.15

$$FP = \frac{\sum_i n_i g_i - 208.30}{2.40 \times \ln(8n_C + 8n_{Si} + 8n_S + 2n_H - 2n_X - 4n_O) + 1} + 196.68$$

$$= \frac{(3 \times 77.8 + 162.43 + 108.68 + 803.82) - 208.30}{2.40 \times \ln(8 \times 5 + 8 \times 0 + 8 \times 0 + 2 \times 12 - 2 \times 0 - 4 \times 1) + 1} + 196.68 = 298.3 \text{ K}$$

1.3 Estimation methods of the FP for mixtures

The values of FP for multicomponent mixtures are also encountered in chemical industries. The FP data for a variety of mixtures are scarce in the literature. Moreover, composition ranges for specific mixtures can vary quite substantially. Thus, experimental determination of the FPs of mixture requires a lot of manpower, time, and material resources with high costs. Development of predictive theoretical models is rather useful and required based on a limited number of initial basic data. For miscible mixtures, some methods have been introduced in the literature for predicting the FPs of different types of mixtures [20, 29, 72–82]. Le Chatelier proposed an empirical method to study firedamp flammability [83]. Mashuga and Crowl [84] proved that the relation of Le Chatelier [83] is theoretically valid where some conditions are satisfied. Liaw et al. [85, 86] have improved the Le Chatelier rule [83] by studying mixtures exhibiting minimum or maximum FPs with the prediction of FPs of miscible and partially miscible mixtures [76, 87] as well as activity coefficients [88]. Gmehling and Rasmussen [75] explored the influence of temperature on flammability limits.

The knowledge of certain characteristic parameters of each compound or mixture, such as reaction enthalpies and activity coefficients, is often required. The calculation of these parameters sometimes may be quite complex and difficult. For those methods where the activity coefficients are needed, the NRTL (Non-Random Two-Liquid), Wilson, and UNIQUAC (UNIversal QUAsiChemical) thermodynamic models are frequently used. Since these models require binary interaction parameters based on experimental data, they are often not used due to the various combinations of possible mixtures. The predictions based on UNIFAC-type models [89, 90] were superior to the NRTL-based or Wilson-based ones [91] for some mixtures because these UNIFAC-based models apply chemical group contributions obtained from a large database to calculate the binary interaction parameters without using any experimental interaction parameters. For further improvement of the performance of the original UNIFAC model, some modifications on this model have also been suggested [91–95]. For example, Liaw and Tsai [96], as well as Liaw and Chen (Liaw and Chen, 2013), have employed the UNIFAC-Dortmund 93 method for predicting the activity coefficients. This approach can provide the most completed database of binary interaction parameters. Moreover, the range of mixtures by application of the modified UNIFAC-Dortmund 93 model in the FP prediction will increase expectably as the numbers of groups and interactions available for the modified UNIFAC-Dortmund 93 increase periodically [96, 97].

Empirical methods have also been developed for predicting the FP of mixtures but they require no a priori knowledge of the property's behavior with respect to its parameters. Many empirical methods involve some form of vapor-liquid equilibrium calculation. They avoid the need for the measured values of physical properties as much as possible. Catoire et al. [19] have developed an equation based on pure compound predictions by involving the vaporization enthalpy, the NBP, and n_C. Later, they extended this equation to mixtures [29, 73]. Pan et al. [98] avoided the need for experimental data of complex flash point or flammability limit of pure components as well as the use of the Le Chatelier law [83]. They employed some commonly used physicochemical properties of chemicals as input parameters to predict the FPs of binary miscible mixtures. Since the FP is related to volatility and thermodynamic properties, the majority of predictive methods for mixtures require accurate values of thermodynamic properties, typically vapor pressure, boiling point, and enthalpy of vaporization. Among different methods, two notable methods of Catoire et al. [73] and Pan et al. [98] based on pure compounds are described here.

1.3.1 Catoire et al. method [73]

For pure compounds, Catoire and Naudet [19] proposed an empirical equation on the basis of approximately 600 compounds for the accurate estimation of the *FP* as:

$$FP = 1.477 \times NBP_{mix}^{0.79686} \times \Delta_{vap}H°(298.15\,K)_{mix}^{0.16845} \times n_{C,mix}^{0.05948} \qquad (1.21)$$

Catoire et al. [29, 73] improved this equation for the fuel mixture. The necessary data for the use of the fuel mixture include NBP_{mix}, $\Delta_{vap}H°(298.15\,K)_{mix}$ and $n_{C,mix}$. The composition of the gas phase must be known in dealing with a fuel mixture here. UNIFAC [75, 90] is a group contribution-based model, which can be used for estimation of liquid-phase activity coefficients γ_i of nonelectrolytic mixtures. Some attempts have been made to improve the predictions of the UNIFAC method [32, 88, 92, 95, 96]. The main drawback is that the UNIFAC method is mathematically uneasy to handle and the calculations are therefore tedious. The values of γ_i coefficients are needed for calculating NBP and $\Delta_{vap}H°(298.15\,\text{K})$ of the liquid mixture.

1.3.1.1 Determination of NBP$_{mix}$

The value of NBP$_{mix}$ is determined according to $\sum_i P_i = 760$ *Torr* where P_i is the partial pressure of compound i in the gas phase above the liquid at $T=$ NBP$_{mix}$. Partial pressures P_i are calculated according to $P_i = x_i\gamma_i P_{isat}$ where x_i is the mole fraction of compound i in the liquid phase $\left(\sum_i x_i = 1\right)$, γ_i is the activity coefficient of compound i in the liquid fuel mixture at NBP$_{mix}$, and P_{isat} is the vapor pressure of pure compound i at NBP$_{mix}$. The values of P_{isat} as a function of T is available either tabulated or expressed according to the Antoine equation [99–102].

1.3.1.2 Determination of $\Delta_{vap}H°(298.15\,\text{K})_{mix}$

The value of $\Delta_{vap}H°(298.15\,\text{K})_{mix}$ is calculated according to the Clausius-Clapeyron law from the slope $\ln P = f(1/T)$, which is equal to $-\Delta_{vap}H°/R$ where $P = \sum_i P_i$ at T. The value of P_i is calculated by the same equations as discussed in the previous section except at T rather than NBP$_{mix}$. Three temperatures in vicinity of 298.15 K, e.g., regression, namely, 293.15, 303.15, and 308.15 K, are considered for this linear $\ln P = f(1/T)$ regression.

1.3.1.2 Determination of $n_{C,mix}$

The value of n_C represents the number of carbon atoms in a fictitious compound representative of the fuel vapor mixture above the fuel liquid mixture. It is defined as $n_{C,mix} = \sum_i n_{C,i}y_i$, where $n_{C,i}$ is the number of carbon atoms in the compound i and y_i is the mole fraction of compound i in the fuel vapor mixture above the liquid at the FP. Since the FP is unknown, y_i is defined here as the mole fraction of compound i in the fuel vapor mixture above the liquid at NBP$_{mix}$.

1.3.2 Pan et al. method [98]

Pan et al. [98] studied the FPs of 28 binary miscible mixtures with different compositions, comprising 18 flammable pure components, which were measured by using the closed cup apparatus. They used these experimental data to develop suitable models for predicting the FPs of binary miscible mixtures. Their models are based on the vapor-liquid equilibrium theory, the NBP, the standard enthalpy of vaporization, the average number of carbon atoms, and the stoichiometric concentration of the gas phase. They used two modeling methods of multiple linear regression (MLR) and multiple non-linear regression (MNR). These models are based on NBP_{mix}, $\Delta_{vap}H°(298.15K)_{mix}$, and $n_{C,mix}$, which have been used by Catoire et al. [29, 73].

1.3.2.1 MLR model
MLR model of Pan et al. [98] for binary miscible mixtures is given as follows:

$$FP = 18.788 + 0.640 \times NBP_{mix} + 1.005 \times \Delta_{vap}H°(298.15\,K)_{mix} + 1.210 \times C_{mix} \quad (1.22)$$
$$- 3.743 \times n_{C,mix}$$

where C_{mix} is the stoichiometric concentration of the combustible components in the vapor phase, which can be calculated using the same method of $n_{C,mix}$. Thus, C_{mix} is defined as $C_{mix} = \sum_i y_i Conc_i$ where $Conc_i$ is the stoichiometric concentration of the combustible component i.

1.3.2.2 MNR model
The second model of Pan et al. [98] for binary miscible mixtures is given as follows:

$$FP = 42.9611 + 0.525 \times NBP_{mix} + 0.833 \times \Delta_{vap}H°(298.15\,K)_{mix}$$
$$+ 1.011 \times C_{mix} - 3.048 \times n_{C,mix} + 0.00049$$
$$\times \left(0.525 \times NBP_{mix} + 0.833 \times \Delta_{vap}H°(298.15\,K)_{mix} + 1.011 \times C_{mix} - 3.048 \times n_{C,mix}\right)^2$$
$$(1.23)$$

The reliability of this model is slightly lower than the MLR model.

1.3.2.3 Comparison of two models of Pan et al. [98] and Catoire et al. [73]
Pan et al. [98] have compared their results with the work of Catoire et al. (2006a). Since these models were developed based on two different datasets, Pan et al. [98] have employed the model of Catoire et al. [73] to predict the FP values for the 73 data points in the same test set as used in their work [98]. For MLR model of Pan et al. [98], the AAE and APE values are 2.506 K and 0.874%, respectively. The values of AAE and APE of MNR model of Pan et al. [98] are also close to MLR model, i.e., 2.537 K and 0.885%, respectively. Meanwhile, AAE and APE for the Catoire's model [73] are 3.246 K and 1.135%, respectively. By comparing the results of the Catoire's model with MLR and MNR models of Pan et al. [98], it can be concluded that the results obtained on the same test set by Pan et al. models [98] are superior to those of the Catoire et al. model [73]. Higher accuracy of Pan et al. [98] models is probably because the four physicochemical parameters employed are likely to have stronger relationships with

the FP property of binary mixtures than the three parameters employed in Catoire et al. model [73]. Thus, the inclusion of the fourth parameter of C_{mix} in Pan et al. models [98] shows that the stoichiometric concentration of the gas phase mixture, which characterized the combustion properties of binary miscible mixtures, had been taken into account for the modeling.

The developed models possess some disadvantages:

(1) They require the values of some other physicochemical parameters the availability or the lack of which may result in some limitations on their applicability range.
(2) The accuracy of these models may be quite dependent on that of the needed physicochemical parameters.
(3) As compared to different models of FP of pure compounds, various available methods for estimation of the FP of mixture require complex calculations and expert users.

For predicting the FP of the mixture, additional works are still necessary and required for the development of more accessible and precise parameters or descriptors capable to correlate the mixture FP.

1.4 Summary

This chapter introduces different methods for predicting the FP of different classes of compounds, which belong to three different categories of pure compounds – empirical, QSPR, and SGC methods. Empirical methods require at least one experimental data such as the NBP. QSPR methods usually need complex molecular descriptors, computer codes, and expert users. Among QSPR methods, several methods are based on simple molecular structural parameters, which have been described in this chapter. For the application of the SGC methods, some of the best among them are also demonstrated. In contrast to pure compounds, available methods are complex for mixtures. The calculations of these methods are difficult for users because they require many parameters for evaluation of the FP of the desired mixture.

2 Autoignition

Autoignition temperature (AIT) is referred to as spontaneous ignition temperature, self-ignition temperature, and autogenous ignition temperature. It is the lowest temperature at which the substance will produce hot-flame ignition in the air at atmospheric pressure without the aid of an external energy source such as spark or flame. For the combustible mixture, it is the lowest temperature in which the rate of heat evolved by the exothermic oxidation reaction will overbalance the rate at which heat is lost to the surroundings and cause ignition. It is an important factor in environmental impact because it has extended in view of the possibility of accidents causing significant damage to people and the environment. An exothermic oxidative reaction can be initiated spontaneously at elevated temperatures where the oxygen in the air can react with the combustible material. Ignition can occur at the AIT because the rate of the heat releasing overbalances the rate at which heat is lost to the surroundings. Moreover, the AIT is one of the most important safety specifications because it is used to characterize the hazard potentials of a chemical substance. For organic materials, the AIT is one of the incidents of fire disasters caused by the leak of material. Different factors can influence the experimental values of AIT, that is, physicochemical properties of the substance, test pressure, oxygen concentration, and the volume and the material of the vessel used in the method. Thus, AIT measurement is laborious and might even be impossible for some hazardous compounds.

It is important to understand the autoignition process for handling, transporting, and storing combustible compounds. Determination of the AIT by experiment is very laborious for those compounds that have hazardous properties such as toxicity [103] and explosivity when exposed to air [104–108] (e.g., rocket liquid fuels). All databases usually reported the AIT values of compounds with no information of experimental method used. For safety purposes, it is also important to have suitable methods for predicting AIT of new or desirable compounds that can reduce costs associated with synthesis, test, and evaluation of the materials. Owing to the importance and different behavior of the AIT of various classes of hydrocarbons, some efforts have been made to introduce suitable methods for prediction of their AIT. The AIT of a hydrocarbon is very dependent upon its molecular structure. It is possible to obtain a general correlation for different classes of hydrocarbons [55, 109, 110]. The AIT of hydrocarbons proceeds by a free radical reaction. The stability of the free radical intermediates can determine the ease of oxidation of desired compound [111]. Different structural features may affect the mechanism of oxidation in autoignition. Valuable insight into the structural significance and the AIT trends of hydrocarbons can determine how susceptible individual structural features are to oxidative attack and autoignition. Various parameters such as chain length, the addition of methyl groups, unsaturation, chain branching, and cyclic and aromatic structures can

https://doi.org/10.1515/9783110572223-002

elevate the AIT values. The AIT decreases as the number of carbon atoms increases for most molecules in homologous series of hydrocarbons. A high degree of branching in a hydrocarbon molecule helps stabilize it against spontaneous ignition. Methylene groups can increase the potential for autoignition that can be averted if branching is used to limit the number of such moieties. For mixtures of different hydrocarbon fuels, it is possible to calculate the AIT from a comprehensive analysis of the individual components in them and their contribution to overall autoignition quality. In this chapter, the method of measurement of the AIT is discussed. Then, simple approaches for prediction of the AIT for pure compounds are demonstrated.

2.1 Measurement of the AIT

The AIT of a specific compound may be reported differently in different literature, up to several hundred Kelvin; experimental data from different sources have used [112]. Fortunately, deviations are large only for little compounds. Most of the experimental data of different compounds are available in DIPR 801 [43] and the chemical database of the department of chemistry at the University of Akron (USA) [70]. DIPR 801 database [43] contains physical properties and flammability characteristics of a large number of pure compounds. The second database provides properties for many hazardous chemicals from a large number of scientific sources.

Experimental determination of the AIT depends on the apparatus used and the test method employed. The measured AIT values presented by different sources can diverge dramatically. Similar to most experimental measurements, obtaining the AIT experimentally is costly in terms of the time and resources required, especially when dealing with materials that are dangerous to handle (e.g., toxic) (Mitchell and Jurs, 1997). Thus, according to the difficulties regarding the measurement of safety properties of materials, reliable theoretical methods are desirable. In addition, once a reliable prediction method is developed, it can be utilized for computer-aided molecular design (CAMD) studies to create new opportunities in the virtual synthesis and evaluation of chemicals.

The value of AIT is required to supply the activation energy needed for combustion. It decreases as the pressure or oxygen concentration increases. For liquid chemicals, the AITs are typically measured using a 500 ml glass flask placed in a temperature-controlled oven in accordance with the procedure described in ASTM E659 [113]. As shown in Figure 2.1, approximately 100 μl of fuel, as defined by ASTM E659, is injected into a uniformly preheated glass flask containing air at a predetermined temperature. When the liquid enters, it evaporates and mixes with the surrounding air.

The AITs for hydrocarbon fuels depend on both the molecular structure of the fuel and the surface material. The measurement of AIT often requires long-exposure times, as much as 10 min, without the capability of varying the material composition

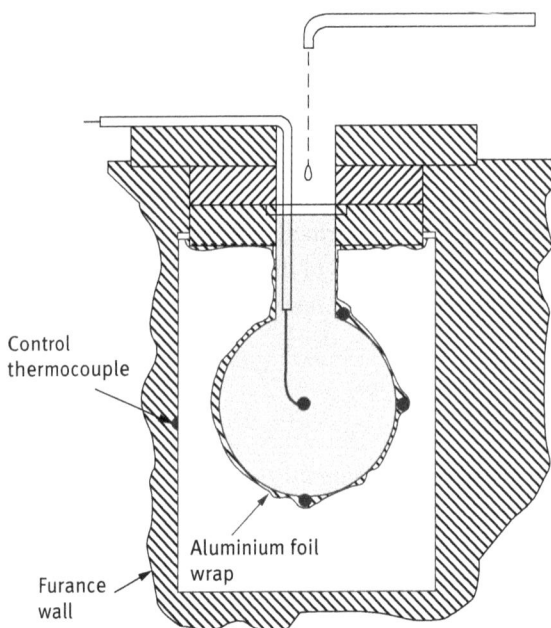

Figure 2.1: Apparatus for measuring the minimum value of AIT of liquid chemicals as defined by ASTM E659.

of the hot surface. Although liquids are commonly injected into the hot flask, the ASTM standard states that no condensed phase, liquid or solid, should be present when ignition occurs. Thus, this test attempts to measure the AIT of hydrocarbon fuels in the gas phase. For long-duration exposures of the fuels to the heated surface, AIT for a complex mixture of the products of fuel decomposition is determined rather than that of the original fuel. If the measurement of the AIT involves short-exposure times, it would greatly reduce the unknown extent of fuel decomposition and provide more meaningful data on fuel structure and surface effects. For short-exposure times, the measurements are also more relevant for determining hazards arising from malfunctions and accidents.

2.2 Predictive methods of the AIT for pure compounds

Neural network, QSPR and SGC are three usual convenient methods, which have been used to predict AIT of hydrocarbons and organic compounds [55, 109, 110, 112, 114–129]. QSPR methods are usually based on complex molecular descriptors, which require special software to obtain unconventional parameters used in the equations of the desired models. The selection of appropriate molecular descriptors such as topological indices, quantum chemical parameters, and electrostatic indices is a key problem in QSPR models. Multiple linear regression (MLR), partial least-squares (PLS) regression, and principal component regression (PCR) as linear models as well as ANN, genetic programming (GP), support vector machines (SVMs), and adaptive neuro-fuzzy inference systems are different types of ANNs, which are usually used for combining the selected molecular descriptors. It was tried to improve QSPR studies for reliable prediction of the AIT by choosing larger dataset. Suitable QSPR models are derived by increasing the square of the correlation coefficient (R^2) and decreasing the rms error of training, validation, and testing sets. For all QSPR methods, it is important to use a large dataset of different molecular structures. Moreover, the compounds with similar molecular structure in the training set of the QSPR procedure should be used as test set. QSPR models on the basis of unusual and complex molecular descriptors require specific computer codes such as Dragon [63] for obtaining descriptors. For example, Suzuki [116] and Pan et al. [121] have used QSPR methods for prediction of the AIT of organic compounds and hydrocarbons, respectively. QSPR model of Suzuki [116] contained unconventional physicochemical parameters such as critical pressure and parachor. Owing to the presence of unusual parameters in Suzuki's model [116], it has some limitations on its application. Pan et al. [121] have also introduced atom-type electrotopological-state indices together with both electronic and topological characteristics of the analyzed molecules for prediction of the AIT of 118 hydrocarbons by means of ANN. A simple QSPR method based on molecular structure descriptors has been introduced for predicting the AIT of different classes of hydrocarbons, which is consistent with different behaviors of free radical for decreasing or increasing of the AIT [112].

SGCs were also used to predict the AIT of organic compounds including hydrocarbons. They directly use the information of the molecular structure to predict the AIT. Since SGCs can be easily applied through desk calculation without using computer codes, it may be more attractive for prediction of the AIT. Albahri [55], Albahri and George [109], and Chen et al. [110] introduced three SGCs for prediction of the AIT of hydrocarbons and organic compounds, respectively. Albahri [55] used the data of 131 pure hydrocarbon fluids on the basis of the AIT to obtain 20 group contributions. The method of Albahri [55] has erroneous predictions for the AIT of long chain hydrocarbons such as tricosane because it estimates negative values of the AIT for them. Albahri and George [109] introduced 58 single and binary structural groups from the AIT of 490 compounds to calculate the AIT of organic compounds containing heteroatoms other than pure hydrocarbons. Their method cannot be used for some alkynes because several structural group contribution values have not been defined. Chen et al. [110] proposed a suitable SGC, which includes 45 molecular functional groups and is a polynomial of degree 3, which was built up using a 400-compound training set. They have tested their model with 83 extra compounds that were not included in the original training set. For long straight chain saturated hydrocarbons, the measured AIT values are nearly constant. Thus, it was shown that application of these SGCs gives large deviations for long-chain saturated hydrocarbons [112]. In contrast to QSPR models on the basis of complex molecular descriptors, SGCs can be easily applied for calculation of the AIT from only the chemical structure of the desired compound. Moreover, SGCs have several shortcomings: (1) they cannot be used for the new compounds with functional groups not included in those used for the model development; (2) they also provide a weak ability in distinguishing the isomeric compounds; and (3) these methods may provide unreliable and large deviations for some compounds. Combination of ANN and SGC (ANN-SGC) has been used to evaluate AIT of 1,025 compounds in three different sets using large structural moieties, that is, 146 functional groups [124]. In contrast to SGCs, ANN-SGC contains too many weighting parameters and requires special software. In this chapter, several simple methods are demonstrated.

2.2.1 The use of SGC by a polynomial of degree 3 for organic compounds

Chen et al. [110] proposed a model for prediction of the AIT of organic compounds based on SGC. This model has been built up using a 400-compound training set with $R^2 = 0.8474$ and an average error of 32 K. The predictive capability of the proposed model has been tested for 83 further compounds. The predictive capability for the second set is about 0.5361 with an average error of 70 K. This model is a polynomial of degree 3 as:

$$\text{AIT} = 750.3 + \left(\sum_i v_i f_i \right) - 8.644 \times 10^{-4} \left(\sum_i v_i f_i \right)^2 - 4.5604 \times 10^{-6} \left(\sum_i v_i f_i \right)^3 \quad (2.1)$$

Chen et al. [110] have compared the predicted results of eq. (2.1) with Albahri and George's work [109]. Equation (2.1) exhibits better performance in terms of R^2, which may be attributed to the modification of the group definitions and not the type of empirical model chosen. The addition of halogen atoms to non-ring hydrocarbons and ring-structure compounds has different effects on their AITs. For the model of Chen et al. [110], 14 new groups were introduced to discriminate the mentioned different effects for halogen compounds.

Example 2.1: Using eq. (2.1), calculate the AIT for the following compounds:

(a) Decahydronaphthalene (the measured AIT = 528 K [70])

(b) Cyclodecane (the measured AIT = 508 K [43])

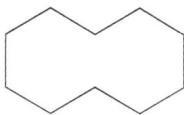

(c) Cyclopentene (the measured AIT = 668.15 K [43])

(d) Naphthalene (the measured AIT = 813.15 K [116])

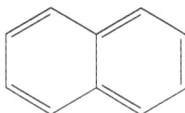

(e) 1,1,2-Trichloroethene (the measured AIT = 693 K [116])

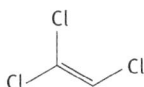

$$Cl$$

$$Cl \diagup \diagdown Cl$$

Answer – Using eq. (2.1) and Table 2.1, we get

a) $\sum_i v_i f_i = v_{> CH_2(ring)} f_{> CH_2(ring)} + v_{> CH-\ (ring)} f_{> CH-\ (ring)}$

$$= 8 \times (-28.4401) + 2 \times (-6.7179) = -240.957$$

$$AIT = 750.3 + \left(\sum_i v_i f_i \right) - 8.644 \times 10^{-4} \left(\sum_i v_i f_i \right)^2 - 4.5604 \times 10^{-6} \left(\sum_i v_i f_i \right)^3$$

$$= 750.3 + (-240.957) - 8.644 \times 10^{-4} (-240.957)^2 - 4.5604 \times 10^{-6} (-240.957)^3$$

$$= 523.0 \text{ K}$$

b) $\sum_i v_i f_i = v_{> CH_2(ring)} f_{> CH_2(ring)} = 10 \times (-28.4401) = -284.401$

$$AIT = 750.3 + \left(\sum_i v_i f_i \right) - 8.644 \times 10^{-4} \left(\sum_i v_i f_i \right)^2 - 4.5604 \times 10^{-6} \left(\sum_i v_i f_i \right)^3$$

$$= 750.3 + (-284.401) - 8.644 \times 10^{-4} (-284.401)^2 - 4.5604 \times 10^{-6} (-284.401)^3$$

$$= 500.9 \text{ K}$$

Table 2.1: The structural group contribution values for estimation.

Serial No.	Group	f_i	Serial No.	Group	f_i
1	–CH₃	–22.8857	25	–F (ring)	88.8289
2	>CH₂	–28.5961	26	–Cl (ring)	79.4122
3	>CH–	1.3340	27	–Br (ring)	53.4870
4	>C<	49.7423	28	–OH (alcohol)	–8.9378
5	=CH₂	–21.6668	29	–OH (phenol)	134.3524
6	=CH–	–46.3286	30	–O– (non-ring)	–70.0383
7	=C<	–32.7605	31	–O– (ring)	–28.4801
8	CH	–81.0169	32	>C=O (non-ring)	8.1173
9	C–	–64.4957	33	>C=O (ring)	57.5044
10	>CH₂ (ring)	–28.4401	34	–HC=O (aldehyde)	–138.3186
11	>CH– (ring)	–6.7179	35	–COOH (acid)	4.0037
12	>C< (ring)	–21.7342	36	–COO–	35.2011
13	=CH– (ring)	19.5293	37	–NH₂	–17.7579
14	=C< (ring)	–49.068	38	>NH (non-ring)	–1.8223
15	=CH–	6.2350	39	>NH (ring)	24.5474
16	>C (fused)	5.8332	40	>N– (non-ring)	–4.7926
17	>C (non-fused)	15.9976	41	>N– (ring)	–49.3834
18	–CH₃ (attached to at least one halogen atom)	103.2738	42	–N= (non-ring)	–41.9897
19	>CH₂ (attached to at least one halogen atom)	–9.8344	43	–N= (ring)	31.8743
20	>CH– (attached to at least one halogen atom)	–24.2759	44	–CN	80.5038
21	>C< (attached to at least one halogen atom)	293.5064	45	–NO₂	–52.7670
22	–F (non-ring)	–45.0477			
23	–Cl (non-ring)	33.9332			
24	–Br (non-ring)	–27.8628			

c) $\sum_i v_i f_i = v_{>CH_2(ring)} f_{>CH_2(ring)} + v_{=CH-\ (ring)} f_{=CH-\ (ring)}$

$$= 3 \times (-28.4401) + 2 \times (19.5293) = -46.2617$$

$$AIT = 750.3 + \left(\sum_i v_i f_i\right) - 8.644 \times 10^{-4} \left(\sum_i v_i f_i\right)^2 - 4.5604 \times 10^{-6} \left(\sum_i v_i f_i\right)^3$$

$$= 750.3 + (-46.2617) - 8.644 \times 10^{-4}(-46.2617)^2 - 4.5604 \times 10^{-6}(-46.2617)^3$$

$$= 702.6 \text{ K}$$

d) $\sum_i v_i f_i = v_{=CH-\,(ring)} f_{=CH-\,(ring)} + v_{=C<\,(ring)} f_{=C<\,(ring)} = 8 \times (19.5293)$

$$+ 2 \times (-49.068) = 58.0984$$

$$AIT = 750.3 + \left(\sum_i v_i f_i \right) - 8.644 \times 10^{-4} \left(\sum_i v_i f_i \right)^2 - 4.5604 \times 10^{-6} \left(\sum_i v_i f_i \right)^3$$

$$= 750.3 + (58.0984) - 8.644 \times 10^{-4} (58.0984)^2 - 4.5604 \times 10^{-6} (58.0984)^3$$

$$= 804.6 \text{ K}$$

e) $\sum_i v_i f_i = v_{-Cl\,(non-ring)} f_{-Cl\,(non-ring)} + v_{=C<} f_{=C<} + v_{=CH-} f_{=CH-}$

$$= 3 \times (33.9332) + 1 \times (-32.7605) + 1 \times (-46.3286) = 22.7105$$

$$AIT = 750.3 + \left(\sum_i v_i f_i \right) - 8.644 \times 10^{-4} \left(\sum_i v_i f_i \right)^2 - 4.5604 \times 10^{-6} \left(\sum_i v_i f_i \right)^3$$

$$= 750.3 + (22.7105) - 8.644 \times 10^{-4} (22.7105)^2 - 4.5604 \times 10^{-6} (22.7105)^3$$

$$= 772.5 \text{ K}$$

2.2.2 A simple QSPR model for various classes of hydrocarbons

A simple model was introduced for estimating the AITs of different classes of hydrocarbons, including alkanes, alkenes, cycloalkanes, cycloalkenes, alkynes, and aromatics. It contains four variables as follows [112]:

$$AIT = 647 + 33.33n_C - 20.79n_H + 58.20F_{SH} + 81.03F_{BH} \qquad (2.2)$$

Two functions F_{SH} and F_{BH} are related to size and branches of different classes of hydrocarbons, respectively. Function F_{SH} can control the rate of changing of the AIT on the basis of n_C and n_H for molecules in homologous series of hydrocarbons. Function F_{BH} can adjust a high degree of branching in a hydrocarbon molecule. Different molecules with the same n_C have a tendency to react by decreasing n_H in the following order:

$$\text{aromatics} < \text{cyclics} < \text{alkenes} < \text{alkanes}$$

This general structural sequence is consistent with the ease of free radical formation of the mentioned hydrocarbons. This situation confirms foundation of previous works [117, 118, 130]: the AIT mechanism proceeds by a free radical reaction. Two functions F_{SH} and F_{BH} can also affect the ease of free radical formation, and consequently of oxidation. Thus, some of the structural features can affect the mechanism of autoignition that includes chain length, the addition of methyl groups, unsaturation, branching, and strain. Two functions F_{SH} and F_{BH} can be specified according to the following situations.

2.2.2.1 F_{SH}

The value of F_{SH} depends on the size of hydrocarbon and has some conditions for each class:

(i) *Saturated acyclic hydrocarbons*: This condition is applied only for linear alkanes. For n-paraffins with $n_C < 5$, $F_{SH} = 5.5 - n_C$. The value of F_{SH} equals -1.0 for $5 \leq n_C \leq 10$. For $11 < n_C \leq 17$, $F_{SH} = 0.0$. For n-alkanes with $n_C > 17$, AIT is constant and its value is 475 K.

(ii) *Saturated cyclic hydrocarbons*: This state is valid only for cyclic compounds without any substituent. For $n_C \leq 5$, $F_{SH} = 5.5 - n_C$. The value F_{SH} equals -2.0 for $n_C \geq 6$.

(iii) *Alkenes*: (a) For the compounds with general formula $CH_2 = CR_1R_2$, two different cases can be considered: (1) If R_1 and R_2 are methyls, F_{SH} equals 2.0. (2) If one of the alkyl groups is methyl and the other one is a nonlinear alkyl group, F_{SH} is 1.0. (b) For $CH_2 = CH(CH_2)_nCH_3$, where $n = 2$ to 8, F_{SH} equals -1.0. For $n > 8$, AIT is constant and its value is 515 K. (c) For linear alkenes with general formula $R_1CH = CHR_2$ except both R_1 and R_2 are $-CH_3$ groups, F_{SH} equals -1.0. (d) For the existence of two double bonds except 1,3-butadiene, F_{SH} is -1.0.

(iv) *Alkynes*: (a) For the existence of triple bond, F_{SH} is -1.5. (b) The value of F_{SH} equals -2.2 for the presence of both triple and double bonds simultaneously.

2.2.2.2 F_{BH}

(i) *Saturated acyclic hydrocarbons*: The value of F_{BH} depends on the number of alkyl substituents ($n_{alkyl\ sub}$) attached to the longest continuous chain of carbon atoms (LCC): (a) If $n_{alkyl\ sub} > 2$, $F_{BH} = 2.0$. (b) If $n_{alkyl\ sub} = 2$, the values of F_{BH} are 2.0 and 0.8 for LCC <5 and LCC≥ 5, respectively. (c) If $n_{alkyl\ sub} = 1$, the values of F_{BH} are 2.0, 0 and -0.5 for LCC<5, $5 \leq$LCC<7 and LCC≥ 7, respectively.

(ii) *Saturated cyclic hydrocarbons*: For the attachment of only one alkyl (except t-butyl) group to the cyclic ring, $F_{BH} = -0.5$.

(iii) *Benzene derivatives*: (a) For the presence of linear alkyl or vinyl group para to the methyl group in disubstituted benzene ring or highly symmetric 1,3,5-trimethyl-benzene, $F_{BH} = 1.25$. (b) The value of F_{BH} equals 1.5 for the attachment of t-butyl and methyl groups simultaneously to a benzene ring. Meanwhile, for the presence of t-butyl group with the other alkyl groups, $F_{BH} = -1.5$. (c) For the presence of two substituents of alkyl or vinyl groups apart from condition (a), the value of F_{BH} equals 0.25 except ethyl vinyl benzene derivatives for which $F_{BH} = -1.0$. For the presence of more than two substituents of alkyl or vinyl groups, the values of F_{BH} are 0.25 and −0.5 for the presence and absence of at least one methyl group as a substituent, respectively.

(iv) *Polycyclic aromatic hydrocarbons*: (a) The values of F_{BH} are −1.0 and −2.0 for the presence of an alkyl group containing less than three and more than two carbons, respectively, as a substituent attached to the aromatic ring. (b) For the compounds having more than two aromatic rings without any substituent, F_{BH} = −1.75.

Example 2.2: Calculate the AIT for 1-eicosene with the following molecular structure:

Answer: Since condition (iii) (b) of F_{SH} has been satisfied, the value AIT is 515 K. The measured AIT is 510 K [70].

2.2.2.3 Comparison of the reliability of this method with SGC methods

Both eq. (2.2) and SGC methods directly use the information of the molecular structure instead of the physical properties, which are used by the other available QSPR approaches, to predict the AIT. These methods can be easily applied through desk calculation without using computer programs and physical properties. They may be more attractive for prediction of the AIT. Moreover, they can be easily applied to those compounds that are new molecules or their physical properties are unavailable. It was shown that Albahri's method [55] has erroneous predictions for the AIT of long-chain hydrocarbons. Equation (2.2) can be easily applied to complex molecular structures of unsaturated hydrocarbons. Table 2.2 compares the calculated AIT of eq. (2.2) and three different group additivity methods of Abahri [55], Albahri and George [109], and Chen et al. [110]. As seen in Table 2.2, the predicted AIT values for docosane and triacont-1-ene by Albahri's model [55] are negative. Albahri and George's work [109] cannot be applied to two compounds, namely dimethyl acetylene and non-1-yne, because the requested structural groups of these compounds have not been defined. Table 2.3 summarizes a comparison of eq. (2.2) with the mentioned SGC methods.

2.2.3 A new and reliable model for prediction of the AIT of organic compounds containing energetic groups

A simple model was introduced for reliable prediction of the AIT of organic compounds containing energetic functional groups such as nitro, nitrate, nitramine, and peroxide [131]. In this QSPR model, only the molecular structures of organic energetic compounds were required without using complex molecular descriptors and

Table 2.2: Comparison of the predicted AITs of eq. (2.2), Albahri [55], Albahri and George [109], and Chen et al. [110] with experimental data.

Structure	Exp	Equation (2.2)	Dev	Albahri	Dev	Albahri – George	Dev	Chen et al.	Dev
	788 [70]	793	4	729	−59	669	−120	703	−85
	693 [70]	718	25	672	−21	627	−66	658	−35
	610 [70]	612	2	653	43	656	46	647	37
	703 [70]	701	−2	673	−30	664	−39	655	−49
	510 [43]	503	−7	527	17	530	20	567	57
	553 [43]	596	43	541	−12	538	−15	574	21
	475 [70]	473	−2	511	36	421	−54	537	62
	475 [43]	475	0	−84		754	279	878	403
	535 [70]	540	5	531	−4	494	−41	552	17
	521 [70]	532	11	503	−18	469	−52	531	10

503 [70]	432	−72	474	−29	403	−101	490	−14
783 [43]	731	−52	714	−69	647	−136	745	−39
596 [43]	568	−28	651	55			573	−23
506 [43]	527	21	473	−33			490	−16
578 [43]	589	11	586	8	546	−32	578	0
552 [43]	523	−29	481	−71	418	−134	516	−36
581 [43]	581	0	548	−33	464	−117	558	−23
524 [43]	515	−9	−4512		2388	1864	2174	1650
531 [70]	590	58	479	−53	436	−95	520	−11
722 [43]	693	−29	741	19	721	−1	718	−4

(continued)

Table 2.2: (continued)

Structure	Exp	Equation (2.2)	Dev	Albahri	Dev	Albahri – George	Dev	Chen et al.	Dev
	729 [43]	697	-32	674	-55	714	-15	717	-13
	700 [43]	709	9	806	106	743	43	735	35
	690 [43]	708	18	707	17	774	84	741	51
	830 [43]	814	11	815	12	804	1	866	63
	635 [43]	650	15	728	93	697	62	685	50
	828 [43]	814	-14	828	0	804	-24	866	38

Table 2.3: Comparison of various aspects of Albahri [55], Albahri and George [109], and Chen et al. [110].

General properties	Equation (2.2)	Albahri	Albahri-George	Chen et al.
Number of applied independent variables	4	20	53	45
Number of training set	248	131	470 different types of pure components (hydrocarbons, alcohols, acids, ethers, etc.)	400 different types of organic compounds (hydrocarbons, alcohols, acids, ethers, etc.)
Number of test set	26	–	20 compounds containing 11 hydrocarbons	83 compounds containing 21 hydrocarbons
Kind of descriptors	The number of carbon and hydrogen atoms as well as two functions related to size and branches of different classes of hydrocarbons	Single structural groups	Single and binary structural groups	Single structural groups
The rms deviation (train and test)	28 and 27	70 and 47[a]	187 and 391	171 and 335

[a]Deviations of the predicted AITs of Albahri's model [55] with negative sign have been deleted for calculation of the rms deviation.

computer codes that would otherwise need expert users. The measured AIT of 45 organic energetic compounds has been used to construct model building on the basis of their molecular structures and compared with the predicted results of the best available group additivity method. The new model has also been checked for nine compounds with unlike and complex molecular structures that give good predictions. The rms deviations of the new model and group additivity method of Chen et al. [110] are 47.45 and 194.25 K, respectively, for 54 compounds (corresponding to 111 data points). The optimized correlation for organic compounds containing energetic groups can be expressed as follows:

$$\text{AIT} = 1095.27 - 5213.22\,(n_C/\text{MW}) - 13218.7\,(n_O/\text{MW}) + 82.1732\text{AIT}^+_{\text{SPG}} - 136.609\text{AIT}^-_{\text{SPG}}$$

$$(2.3)$$

where MW is the molecular weight of a desired organic compound; $\text{AIT}^+_{\text{SPG}}$ and $\text{AIT}^-_{\text{SPG}}$ are two correcting functions that are related to the presence of specific polar energetic

Table 2.4: Contribution of AIT^+_{SFG} and AIT^-_{SFG}.

Organic compounds containing energetic groups	AIT^+_{SFG}	AIT^-_{SFG}	Condition
RNO_2	1.0	0	Nitroalkanes containing less than four carbon atoms as well as mononitro derivatives
$R-NNO_2$	0	1.5	–
Nitrobenzene	1.5	0	The attachment of only nitro groups or both nitro and amino groups to benzene ring
$RONO_2$	0	1.0	Mononitrate derivatives
$ROOH$	0	1.5	The attachment of –OOH to tertiary carbon atom
	2	0	–
$Ar-OOH$	0	1.0	–
	0	2.0	–
	0	1.5	–

groups under certain conditions. AIT^+_{SPG} and AIT^-_{SPG} can adjust larger and lower high deviations of the predicted AIT, respectively, on the basis of two ratios of n_C/MW and n_O/MW from experimental data. The values of AIT^+_{SPG} and AIT^-_{SPG} are given in Table 2.4. Since the coefficients n_C/MW and n_O/MW have the negative values in eq. (2.3), increasing the values of n_C and n_O as well as decreasing MW can decrease the value of the AIT. For decreasing the AIT of organic energetic compounds, the contribution of AIT^-_{SPG} is another way. Thus, more precautions should be considered for large values of n_C/MW and n_O/MW in addition to the presence of AIT^-_{SPG}. As seen in Table 2.4, the contribution of AIT^-_{SPG} should be considered for cyclic and acyclic nitramines, which is consistent with the relatively high sensitivity of these compounds with respect to impact [132], shock [133], and electric spark [134].

Example 2.3: Calculate the AIT for ethaneperoxoic acid (peracetic acid) with the following molecular structure:

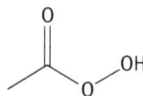

Answer: Using eq. (2.3), we get

$$AIT = 1,095.27 - 5,213.22\,(n_C/MW) - 13218.7\,(n_O/MW) + 82.1732\,AIT_{SPG}^{+}$$
$$- 136.609\,AIT_{SPG}^{-} = 1,095.27 - 5,213.22\,(2/76.05) - 13218.7\,(3/76.05)$$
$$+ 82.1732(0) - 136.609(0) = 436.28\,K$$

Three different values of the AIT were reported for ethaneperoxoic acid as 383.15 [135], 471.15 [136], and 473.15 [137].

2.3 Summary

Since experimental values of the AIT of different classes of compounds are scarce and expensive, this chapter introduces different approaches for prediction of the AIT of some classes of materials. This chapter introduces two different approaches of SGC and QSPR methods for prediction of the AIT of organic compounds and different types of hydrocarbons. Among available SGC and QSPR methods, the method of Chen et al. [110] and eqs (2.2) and (2.3) were demonstrated because they can be used for a wide range of different categories of pure compounds. Available QSPR methods require complex descriptors, computer codes, and expert users, except for eqs (2.2) and (2.3) that are based on molecular structures of hydrocarbons. Equation (2.2) can be used to predict the AITs of different classes of hydrocarbons containing alkanes, alkenes, cycloalkanes, cycloalkenes, alkynes, and aromatics. An elemental composition, as well as two functions FSH and FBH in eq. (2.2), can be easily found from molecular structures of hydrocarbons. As compared to three of the best available SGC methods, that is, Albahri [55], Albahri and George [109], and Chen et al. [110], eq. (2.2) provides more reliable results for different types of hydrocarbons. Equation (2.3) can also be used for prediction of the AIT of organic energetic compounds containing functional groups nitro, nitrate, nitramine, and peroxide.

3 Flammability Limit

It is suitable to keep a combustible compound outside of its flammable concentration range. As mentioned in Chapter 1, the flash point can suggest an approximation of the lower temperature limit in which a chemical evolves enough vapors to support combustion. Flammability limits, which are also referred to as the explosive limits, represent the concentrations of fuel in air, which can support flame propagation. They give the range of fuel concentration usually in terms of percentage volume in the air usually at 298 K where a gaseous mixture can ignite and burn. They are better descriptors of a chemical's flammability in solids, liquids, and gases. A considerable amount of flammability limit data has been published, but the temperature dependence of the limits is nearly always neglected. Available data are frequently reported at 298 K for gases. For liquids and solids, flammability limit data are often reported at a single arbitrary temperature.

The upper and lower flammability limits are some of the most important for safety considerations in storage, processing, and handling, which should be considered in assessing the overall flammability hazard potential of a chemical substance. They are defined in terms of the lower flammability limit (LFL) and the upper flammability limit (UFL) as the degree of susceptibility to ignition or release of energy under varying environmental conditions. Thus, the LFL and UFL show the minimum and maximum concentrations of a combustible substance that is capable of propagating a flame in a homogeneous mixture of the combustible and a gaseous oxidizer (air) under the specified conditions of the test. Below the LFL, there is not enough fuel to cause ignition. If fuel concentration is greater than the UFL, there is insufficient oxygen for the reaction to propagate away from the source of ignition. Determination of the LFL and UFL depends on several variables, e.g., the type of the fuel or chemical, the geometry of the apparatus, strength of the ignition source, test pressure, degree of mixing, oxygen concentration, and concentration of diluents [138].

3.1 Measurement of the LFL and UFL

Differences in apparatuses and experimental methods can influence the measured flammability limits significantly. It is important to use a suitable procedure for the precise determination of the flammability limits. This requirement can be established by the use of a standardized apparatus and conditions as specified in ASTM standard E681 [139], which is shown in Figure 3.1. The values of the LFL and UFL are determined by igniting a uniform mixture of a gas or vapor with air in a closed vessel. In the ASTM standard test method, the upward and outward propagation of the flame, away from the ignition source, is noted by visual

https://doi.org/10.1515/9783110572223-003

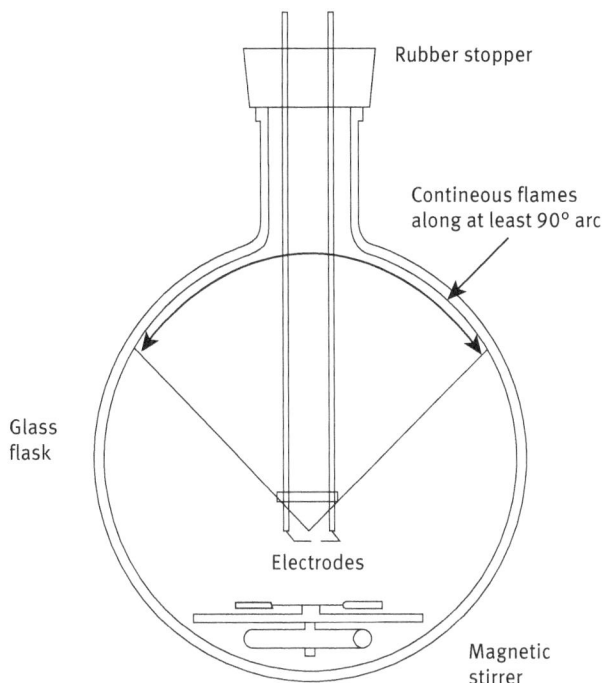

Figure 3.1: Visual criterion for flask flame propagation (ASTM E 681) [139].

observation. The concentration of the desired compound is varied between trials. The composition is determined where propagation of the flame is just sustained. The LFL and UFL may be used to determine guidelines for the safe handling of volatile chemicals. They assess ventilation requirements for the handling of gases and vapors.

3.2 Predictive methods of the flammability limits

Determination of the flammability limits is difficult and not always practical. Thus, a reliable predictive method that is desirably convenient and fast must be used to estimate them. The LFL and UFL are two of the macroscopic properties of compounds, which are related to the molecular structure. The magnitude and predominant types of the intermolecular forces depend on the molecular structure of the desired material. Four different approaches – empirical methods, QSPR based on complex molecular descriptors, ANN-SGC, and SGC – were usually used for estimation of the LFL and UFL [20, 55, 140–177]. The method of SGC suggests that a macroscopic property can be calculated from group contributions. It was indicated that the SGC approach based on non-linear regression and least square techniques suffers from several shortcomings because these limitations are mainly associated by using a simple correlation, which is unable to capture the complex nature of the LFL and UFL properties. Between two properties of the LFL and UFL, the LFL is more important than the UFL. Because of the complex dependency of the LFL on the molecular structure of the compound, it is a difficult property to estimate or correlate. For the coexistence of several functional groups in one molecule, it is difficult to formulate a model that can incorporate the behavior of all the different groups without taking into account the structure of the molecules. For designing safe chemical and petrochemical processes, accurate knowledge of the LFL for a variety of chemicals is needed. The measured LFL and UFL are little for a wide range of chemicals. Moreover, experimental values of the LFL and UFL are rare for most chemicals at non-ambient conditions. For pure chemicals at a single temperature point, usually 298 K, many methods have been developed to estimate their LFL and UFL [20, 55, 140–176]. Vidal et al. [20] reviewed a few of these methods. Available QSPR methods and ANN-SGC are complicated methods, which require computer codes and expert users. In this chapter, the simplest approach for estimating the temperature dependence of the LFL of general organic compounds as well as several simple and reliable SGC models for estimation of the LFL and UFL are discussed.

3.2.1 The predicted LFL as a function of temperature

Catoire and Naudet [164] proposed a simple correlation for the accurate estimation of the LFL of CHNO, and monohalogenated organic pure compounds in air at atmospheric pressure in the 25–400 °C temperature range as:

$$\text{LFL (mol\%)} = 519.957 \times (1 + 5n_C + 1.25n_H - 2.5n_O)^{-0.70936} \times n_C^{-0.197} \times T^{-0.51536} \quad (3.1)$$

where LFL (mol%) is mole percentage and T is the temperature in K.

Example 3.1: Calculate the LFL (mol%) of (a) 2-nitropropane at 25 °C; (b) acetone at 25 °C, and (c) aniline at 140 °C.

Answer: (a) The molecular formula of 2-nitropropane is $C_3H_7NO_2$. Thus, eq. (3.1) gives:

$$\text{LFL (mol\%)} = 519.957 \times (1 + 5n_C + 1.25n_H - 2.5n_O)^{-0.70936} \times n_C^{-0.197} \times T^{-0.51536}$$
$$= 519.957 \times (1 + 5 \times 3 + 1.25 \times 7 - 2.5 \times 2)^{-0.70936} \times (3)^{-0.197}$$
$$\times (298.15)^{-0.51536}$$
$$= 2.70$$

(b) The molecular formula of acetone is C_3H_6O. Thus, eq. (3.1) gives:

$$\text{LFL (mol\%)} = 519.957 \times (1 + 5n_C + 1.25n_H - 2.5n_O)^{-0.70936} \times n_C^{-0.197} \times T^{-0.51536}$$

$$= 519.957 \times (1 + 5 \times 3 + 1.25 \times 6 - 2.5 \times 1)^{-0.70936} \times (3)^{-0.197}$$

$$\times (298.15)^{-0.51536}$$

$$= 2.59$$

(c) The molecular formula of aniline is C_6H_7N. Thus, eq. (3.1) gives:

$$\text{LFL (mol\%)} = 519.957 \times (1 + 5n_C + 1.25n_H - 2.5n_O)^{-0.70936} \times n_C^{-0.197} \times T^{-0.51536}$$

$$= 519.957 \times (1 + 5 \times 6 + 1.25 \times 7 - 2.5 \times 0)^{-0.70936} \times (6)^{-0.197} \times (413.15)^{-0.51536}$$

$$= 1.21$$

The experimental values of 2-nitropropane, acetone, and aniline at the specified temperatures are 2.5, 2.6, and 1.2, respectively [178].

3.2.2 The use of SGC method for prediction of the LFL and UFL of pure hydrocarbons

Albahri [55] presented a suitable SGC method on the basis of 19 structural groups for predicting the LFL and UFL volume %, denoted as the LFL (vol%) and UFL (vol%), respectively, in the air of about 500 pure hydrocarbons. His model has average errors of 0.04 and 1.25 volume % for the LFL (vol%) and UFL (vol%), respectively. He compared the predicted results to that of other methods in the literature and found it to be far more accurate. Table 3.1 shows structural groups for estimating the LFL and UFL volume percent in the air of pure compounds at 298 K. Thus, the LFL (vol%) and UFL (vol%) in the air are calculated by:

$$\text{LFL (vol\%)} = \begin{bmatrix} 4.174 + 0.8093 \times \sum_i (\text{LFL})_i + 0.0689 \times \left(\sum_i (\text{LFL})_i\right)^2 \\ + 0.00265 \times \left(\sum_i (\text{LFL})_i\right)^3 + 3.76 \times 10^{-5} \times \left(\sum_i (\text{LFL})_i\right)^4 \end{bmatrix} \quad (3.2)$$

$$\text{UFL (vol\%)} = \begin{bmatrix} 18.14 + 3.4135 \times \sum_i (\text{UFL})_i + 0.3587 \times \left(\sum_i (\text{UFL})_i\right)^2 \\ + 0.01747 \times \left(\sum_i (\text{UFL})_i\right)^3 + 3.403 \times 10^{-4} \times \left(\sum_i (\text{UFL})_i\right)^4 \end{bmatrix} \quad (3.3)$$

Example 3.2: Calculate the LFL (vol%) of (a) 2,2-dimethylbutane (neohexane); (b) *trans*-1,3-pentadiene; and (c) 1-methyl-2-n-propylbenzene

Table 3.1: Group contribution for estimating the LFL (vol%) and UFL (vol%) in air of pure hydrocarbons at 298 K.

Serial No.	Group	$(LFL)_i$	$(UFL)_i$
1	$-CH_3$ (Paraffin)	-1.4407	-0.8394
2	$>CH_2$ (Paraffin)	-0.8736	-1.1219
3	$>CH-$ (Paraffin)	-0.2925	-1.2598
4	$>C<$ (Paraffin)	0.2747	-2.1941
5	$=CH_2$ (Olefin)	-1.3126	0.2479
6	$=CH-$ (Olefin)	-0.7679	-0.3016
7	$=C<$ (Olefin)	-0.2016	-0.6524
8	$=C=$ (Olefin)	-0.4473	0.0675
9	$\equiv CH$ (Olefin)	-1.2849	3.8518
10	$\equiv C-$ (Olefin)	-0.4396	1.3924
11	$>CH_2$ (Cyclic)	-1.0035	-0.8386
12	$>CH-$ (Cyclic)	-0.4955	-0.9648
13	$>C<$ (Cyclic)	0.1058	-2.2754
14	$=CH-$ (Cyclic)	-0.8700	-0.0821
15	$=C<$ (Cyclic)	-0.5283	-0.1252
16	$=CH-$ (Aromatic)	-0.8891	-1.2966
17	$>CH_2$ (Aromatic)	-1.0884	-1.6166
18	$=C<$ (Aromatic, fused)	-0.3694	-1.4722
19	$=C<$ (Aromatic, non-fused)	-0.2847	0.6649

Answer: (a) 2,2-Dimethylbutane has the following molecular structure:

It has four $-CH_3$ (Paraffin), one $>CH_2$ (Paraffin), and one $>C<$ (Paraffin) groups. The use of eq. (3.2) gives:

$$\sum_i (LFL)_i = 4 \times (-CH_3) + 1 \times (>CH_2) + 1 \times (>C<)$$

$$= 4 \times (-1.4407) + 1 \times (-0.8736) + 1 \times (0.2747) = -6.3617$$

$$LFL\ (vol\%) = \begin{bmatrix} 4.174 + 0.8093 \times \sum_i (LFL)_i + 0.0689 \times \left(\sum_i (LFL)_i \right)^2 \\ + 0.00265 \times \left(\sum_i (LFL)_i \right)^3 + 3.76 \times 10^{-5} \times \left(\sum_i (LFL)_i \right)^4 \end{bmatrix}$$

$$= \begin{bmatrix} 4.174 + 0.8093 \times (-6.3617) + 0.0689 \times (-6.3617)^2 \\ + 0.00265 \times (-6.3617)^3 + 3.76 \times 10^{-5} \times (-6.3617)^4 \end{bmatrix} = 1.19$$

(b) *trans*-1,3-Pentadiene has the following molecular structure:

It has one $-CH_3$ (Paraffin), three $=CH-$ (Olefin), and one $=CH_2$ (Olefin) groups. The use of eq. (3.2) gives:

$$\sum_i (LFL)_i = 1 \times (-CH_3) + 3 \times (=CH-) + 1 \times (=CH_2)$$

$$= 1 \times (-1.4407) + 3 \times (-0.7679) + 1 \times (-1.3126) = -5.057$$

$$LFL\ (vol\%) = \left[\begin{array}{l} 4.174 + 0.8093 \times \sum_i (LFL)_i + 0.0689 \times \left(\sum_i (LFL)_i \right)^2 \\ + \ 0.00265 \times \left(\sum_i (LFL)_i \right)^3 + 3.76 \times 10^{-5} \times \left(\sum_i (LFL)_i \right)^4 \end{array} \right]$$

$$= \left[\begin{array}{l} 4.174 + 0.8093 \times (-5.057) + 0.0689 \times (-5.057)^2 \\ + \ 0.00265 \times (-5.057)^3 + 3.76 \times 10^{-5} \times (-5.057)^4 \end{array} \right] = 1.53$$

(c) 1-Methyl-2-n-propylbenzene has the following molecular structure:

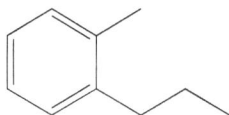

It has two $-CH_3$ (Paraffin), four $=CH-$ (Aromatic), two $=C<$ (Aromatic, non-fused), and two $>CH_2$ (Paraffin) groups. The use of eq. (3.2) gives:

$$\sum_i (LFL)_i = 2 \times (-CH_3) + 3 \times (=CH-) + 2 \times (=C<) + 2 \times (>CH_2)$$

$$= 2 \times (-1.4407) + 3 \times (-0.8891) + 2 \times (-0.2847) + 2 \times (-0.8736) = -8.7544$$

$$LFL\ (vol\%) = \left[\begin{array}{l} 4.174 + 0.8093 \times \sum_i (LFL)_i + 0.0689 \times \left(\sum_i (LFL)_i \right)^2 \\ + 0.00265 \times \left(\sum_i (LFL)_i \right)^3 + 3.76 \times 10^{-5} \times \left(\sum_i (LFL)_i \right)^4 \end{array} \right]$$

$$= \left[\begin{array}{l} 4.174 + 0.8093 \times (-8.7544) + 0.0689 \times (-8.7544)^2 \\ + 0.00265 \times (-8.7544)^3 + 3.76 \times 10^{-5} \times (-8.7544)^4 \end{array} \right] = 0.81$$

The experimental values of 2,2-dimethylbutane (neohexane), *trans*-1,3-pentadiene, and 1-methyl-2-n-propylbenzene are 1.2, 1.52, and 0.82, respectively [55].

Example 3.3: Calculate the UFL (vol%) of (a) *cis*-1,2-dimethylcyclopentane; (b) ethylacetylene (1-butyne); and (c) 1,1-diphenyltetradecane.

Answer: (a) *cis*-1,2-Dimethylcyclopentane has the following molecular structure:

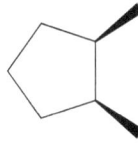

It has two $-CH_3$ (Paraffin), three $>CH_2$ (Cyclic), and two $>CH-$ (Cyclic) groups. The use of eq. (3.3) gives:

$$\sum_i (UFL)_i = 2 \times (-CH_3) + 3 \times (>CH_2) + 2 \times (>CH-)$$

$$= 2 \times (-0.8394) + 3 \times (-0.8386) + 2 \times (-0.9648) = -6.1242$$

$$UFL \ (vol\%) = \left[\begin{array}{l} 18.14 + 3.4135 \times \sum_i (UFL)_i + 0.3587 \times \left(\sum_i (UFL)_i \right)^2 \\ + \ 0.01747 \times \left(\sum_i (UFL)_i \right)^3 + 3.403 \times 10^{-4} \times \left(\sum_i (UFL)_i \right)^4 \end{array} \right]$$

$$= \left[\begin{array}{l} 18.14 + 3.4135 \times (-6.1242) + 0.3587 \times (-6.1242)^2 \\ + \ 0.01747 \times (-6.1242)^3 + 3.403 \times 10^{-4} \times (-6.1242)^4 \end{array} \right] = 7.15$$

(b) Ethylacetylene (1-butyne) has the following molecular structure:

It has one $-CH_3$ (Paraffin), one $>CH_2$ (Paraffin), one $\equiv CH$ (Olefin) and one $\equiv C-$ (Olefin) groups. The use of eq. (3.3) gives:

$$\sum_i (UFL)_i = 1 \times (-CH_3) + 1 \times (>CH_2) + 1 \times (\equiv CH) + 1 \times (\equiv C-)$$

$$= 1 \times (-0.8394) + 1 \times (-1.1219) + 1 \times (3.8518) + 1 \times (1.3924) = 3.2829$$

$$UFL \ (vol\%) = \left[\begin{array}{l} 18.14 + 3.4135 \times \sum_i (UFL)_i + 0.3587 \times \left(\sum_i (UFL)_i \right)^2 \\ + \ 0.01747 \times \left(\sum_i (UFL)_i \right)^3 + 3.403 \times 10^{-4} \times \left(\sum_i (UFL)_i \right)^4 \end{array} \right]$$

$$= \left[\begin{array}{l} 18.14 + 3.4135 \times (3.2829) + 0.3587 \times (3.2829)^2 \\ + \ 0.01747 \times (3.2829)^3 + 3.403 \times 10^{-4} \times (3.2829)^4 \end{array} \right] = 33.87$$

(c) 1,1-Diphenyltetradecane has the following molecular structure:

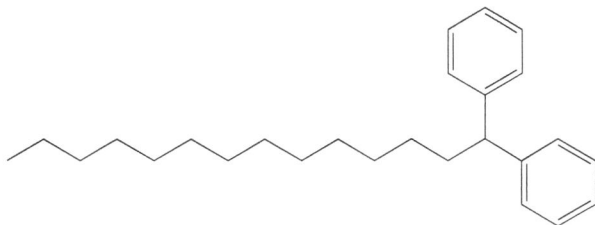

It has one −CH$_3$ (Paraffin), ten =CH− (Aromatic), two =C< (Aromatic, non-fused), twelve >CH$_2$ (Paraffin), and one >CH− (Paraffin) groups. The use of eq. (3.3) gives:

$$\sum_i (\text{UFL})_i = 1 \times (-\text{CH}_3) + 10 \times (=\text{CH}-) + 2 \times (=\text{C}<) + 12 \times (>\text{CH}_2-) + 1 \times (>\text{CH}-)$$

$$= 1 \times (-0.8394) + 10 \times (-1.2966) + 2 \times (0.6649) + 12 \times (-1.1219)$$
$$+ 1 \times (-1.2598) = -27.20$$

$$\text{UFL (vol\%)} = \begin{bmatrix} 18.14 + 3.4135 \times \sum_i (\text{UFL})_i + 0.3587 \times \left(\sum_i (\text{UFL})_i \right)^2 \\ + 0.01747 \times \left(\sum_i (\text{UFL})_i \right)^3 + 3.403 \times 10^{-4} \times \left(\sum_i (\text{UFL})_i \right)^4 \end{bmatrix}$$

$$= \begin{bmatrix} 18.14 + 3.4135 \times (-27.20) + 0.3587 \times (-27.20)^2 \\ + 0.01747 \times (-27.20)^3 + 3.403 \times 10^{-4} \times (-27.20)^4 \end{bmatrix} = 25.37$$

The experimental values of *cis*-1,2-dimethylcyclopentane, ethylacetylene (1-butyne), and 1,1-diphenyltetradecane are 7.3, 32.93, and 24.13, respectively [55].

3.2.3 Extended method for prediction of the UFL of pure compounds

Albahri [179] has extended eq. (3.3) for prediction of the UFL of pure compounds. He introduced 30 atom-type structure groups to represent the UFL for 550 pure substances. His model has a correlation coefficient of 0.9996 and average absolute deviation of 0.17 vol% as:

$$\text{UFL (vol\%)} = \begin{bmatrix} 3.563 + 0.5237 \times \sum_i (\text{UFL})_i + 1.572 \times 10^{-9} \times \left(\sum_i (\text{UFL})_i \right)^2 \\ + 6.375 \times 10^{-8} \times \left(\sum_i (\text{UFL})_i \right)^3 + 3.266 \times 10^{-5} \times \left(\sum_i (\text{UFL})_i \right)^4 \end{bmatrix}$$

(3.4)

Table 3.2 shows structural groups for estimating the UFL volume percent in the air of pure compounds at 298 K.

Example 3.4: Calculate the UFL (vol%) of (a) cyclopentanone, (b) 1-decanol, and (c) n-decanoic acid.

Table 3.2: Group contribution for estimating the UFL (vol%) in air of pure compounds at 25°C.

Serial No.	Group	(UFL)$_i$
1	$-CH_3$	1.114692
2	$>CH_2$	0.339248
3	$>CH-$	0.138901
4	$>C<$	-2.50000
5	$=CH_2$	2.305876
6	$=CH-$	2.024712
7	$=C<$	1.346964
8	$=C=$	6.542315
9	$\equiv CH$	18.1876
10	$\equiv C-$	2.243466
11	$>CH_2$ (Ring)	0.570317
12	$>CH-$ (Ring)	0.70827
13	$>C<$ (Ring)	0.403864
14	$=CH-$ (Ring)	0.672134
15	$=C<$ (Ring)	0.654589
16	$-F$	4.231781
17	$-Cl$	2.254908
18	$-OH$ (Alcohol)	6.41015
19	$-O-$ (Non-ring)	13.04636
20	$>C=O$ (Non-ring) Ketone	5.417596
21	$O=CH-$ (Aldehyde)	13.023
22	$-COOH$ (Acid)	0.047503
23	$=O$	0.001000
24	$-NH_2$	8.603798
25	$>N-$ (Non-ring)	1.900000
26	$-O-$ (Ring)	3.179851
27	$>C=O$ (Ring)	9.414214
28	$>NH$ (Ring)	6.019325
29	$-N=$ (Ring)	2.458781
30	$-H$	12.51823

Answer: (a) Cyclopentanone has the following molecular structure:

It has four $>CH_2$ (Ring) and one $>C=O$ (Ring) groups. The use of eq. (3.4) gives:

$$\sum_i (UFL)_i = 4 \times (>CH_2) + 1 \times (>C=O)$$

$$= 4 \times (0.570317) + 1 \times (9.414214) = 11.69548$$

$$UFL\ (vol\%) = \left[\begin{array}{l} 3.563 + 0.5237 \times \sum_i (UFL)_i + 1.572 \times 10^{-9} \times \left(\sum_i (UFL)_i \right)^2 \\ + \ 6.375 \times 10^{-8} \times \left(\sum_i (UFL)_i \right)^3 + 3.266 \times 10^{-5} \times \left(\sum_i (UFL)_i \right)^4 \end{array} \right]$$

$$= \left[\begin{array}{l} 3.563 + 0.5237 \times 11.69548 + 1.572 \times 10^{-9} \times (11.69548)^2 \\ + \ 6.375 \times 10^{-8} \times (11.69548)^3 + 3.266 \times 10^{-5} \times (11.69548)^4 \end{array} \right] = 10.30$$

(b) 1-Decanol has the following molecular structure:

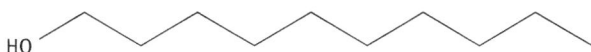

It has one $-CH_3$, nine $>CH_2$, and one $-OH$ (Alcohol) groups. The use of eq. (3.4) gives:

$$\sum_i (UFL)_i = 1 \times (-CH_3) + 9 \times (>CH_2) + 1 \times (-OH)$$

$$= 1 \times (1.114692) + 9 \times (0.570317) + 1 \times (6.41015) = 10.57807$$

$$UFL\ (vol\%) = \left[\begin{array}{l} 3.563 + 0.5237 \times \sum_i (UFL)_i + 1.572 \times 10^{-9} \times \left(\sum_i (UFL)_i \right)^2 \\ + \ 6.375 \times 10^{-8} \times \left(\sum_i (UFL)_i \right)^3 + 3.266 \times 10^{-5} \times \left(\sum_i (UFL)_i \right)^4 \end{array} \right]$$

$$= \left[\begin{array}{l} 3.563 + 0.5237 \times 10.57807 + 1.572 \times 10^{-9} \times (10.57807)^2 \\ + \ 6.375 \times 10^{-8} \times (10.57807)^3 + 3.266 \times 10^{-5} \times (10.57807)^4 \end{array} \right] = 9.51$$

(c) n-Decanoic acid has the following molecular structure:

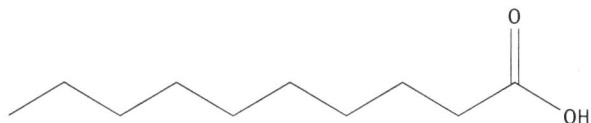

It has one $-CH_3$, eight $>CH_2$, and one $-COOH$ (Acid) groups. The use of eq. (3.4) gives:

$$\sum_i (UFL)_i = 1 \times (-CH_3) + 8 \times (>CH_2) + 1 \times (-COOH)$$

$$= 1 \times (1.114692) + 8 \times (0.570317) + 1 \times (0.047503) = 3.876179$$

$$
\text{UFL (vol\%)} = \left[
\begin{array}{l}
3.563 + 0.5237 \times \sum_i (UFL)_i + 1.572 \times 10^{-9} \times \left(\sum_i (UFL)_i \right)^2 \\
+ 6.375 \times 10^{-8} \times \left(\sum_i (UFL)_i \right)^3 + 3.266 \times 10^{-5} \times \left(\sum_i (UFL)_i \right)^4
\end{array}
\right]
$$

$$
= \left[
\begin{array}{l}
3.563 + 0.5237 \times 3.876179 + 1.572 \times 10^{-9} \times (3.876179)^2 \\
+ 6.375 \times 10^{-8} \times (3.876179)^3 + 3.266 \times 10^{-5} \times (3.876179)^4
\end{array}
\right] = 5.60
$$

The experimental values of cyclopentanone, 1-decanol, and n-decanoic acid are 10.4, 5.5, and 5.5, respectively [43].

3.3 Summary

This chapter introduces different approaches for prediction of the LFL and UFL of pure hydrocarbons as well as some classes of pure compounds. Among three different methods – ANN-SGC, QSPR, and SGC methods – available ANN-SGC and QSPR methods usually need complex molecular descriptors, computer codes, and the expert users. Thus, several simple and reliable SGC methods have been described in this chapter. Equation (3.1) provides a simple path for the accurate estimation of the LFL (mol%) of CHNO, and monohalogenated organic pure compounds in air at atmospheric pressure in the 25–400 °C temperature range. Equations (3.2) and (3.3) give simple correlations for calculating the LFL (vol%) and UFL (vol%) of pure hydrocarbons in the air at 298 K. Finally, eq. (3.4) is an improved correlation, which can be used as pure compounds.

4 Heat of Combustion

The heat of combustion of a specified substance can be defined as the heat evolved when it is converted to the standard oxidation products by means of molecular oxygen [148]. It can be used for reactive materials to estimate the potential fire hazards of chemicals once they ignite and burn. The fuel can be either liquid or solid, which contains only the elements carbon, hydrogen, nitrogen, and sulfur. For complete combustion of the fuel, the products of combustion, in oxygen, are gaseous carbon dioxide, nitrogen oxides, sulfur dioxide, and liquid or gaseous water. There are two different values of specific heat energy for the same batch of combustible organic compounds, i.e. the gross (or high) heat of combustion and the net (or low) heat of combustion. The gross heat of combustion is the quantity of energy released when a unit mass of a combustible compound is burned in a constant volume enclosure, with the products being gaseous, other than water that is condensed to the liquid state. The net (or low) heat of combustion is the quantity of energy released when a unit mass of fuel is burned in a constant pressure, with all of the products, including water, being gaseous. The difference between the gross and net heat of combustion is significant, about 8% or 9%. Because engines exhaust water as a gas, the net heat of combustion is the appropriate value to use for comparing fuels. For a diesel fuel, the net heat of combustion corresponds to a heating value in which the water remains a vapor and does not yield its heat of vaporization. Thus, the energy difference between the gross and net heat of combustion is due to the heat of vaporization of water as [180]:

$$\Delta H_c^o(\text{gross}) = \Delta H_c^o(\text{net}) + \frac{m_{\text{water}}}{m_{\text{diesel}}} \times 40.80 \text{ kJ mol}^{-1} \tag{4.1}$$

where $\Delta H_c^o(\text{gross})$ and $\Delta H_c^o(\text{net})$ are the gross and net heat of combustion, respectively; m_{water} and m_{diesel} are the mass of liquid water in the combustion products and the mass diesel fuel, respectively.

The net heat of combustion can be defined as the increase in enthalpy when a substance containing carbon, hydrogen, nitrogen, oxygen, fluorine, bromine, iodine, sulfur, phosphorous, and silicon atoms in its standard state (298.15 K and 1 atm) undergoes an oxidization to produce its final combustion products: CO_2 (g), F_2 (g), Cl_2 (g), Br_2 (g), I_2 (g), SO_2 (g), N_2 (g), H_3PO_4 (s), H_2O (g), and SiO_2 (cristobalite) [43]. It is a measure of the energy available from a fuel, which is used to compare the heating values of fuels and the stability of compounds. It can be used to assess the potential fire hazard of reactive chemicals and predict the performance of explosive and propellant formulations. The knowledge of $\Delta H_c^o(\text{net})$ for chemicals is essential when considering the thermal efficiency of equipment used to produce power or heat. It also provides a good assessment of the environmental impact of any plant at which complete and incomplete combustion are yet to be defined. For pure chemical

https://doi.org/10.1515/9783110572223-004

compounds, the values of $\Delta H_c^o(\text{net})$ are compiled in databases such as AIChE-DIPPR [43] and API-TDB [181]. These databases contain various chemical families of organic compounds including halogenated compounds, acids, ethers, ketones, aldehydes, alcohols, phenols, esters, amines, anhydrides, and sulfur compounds.

Experimental determination of $\Delta H_c^o(\text{net})$ is tedious, expensive and sometimes impossible. A fast, easy, and accurate estimation method can be useful if experimental values of $\Delta H_c^o(\text{net})$ are not available and determining them experimentally is inconvenient or not possible. The predicted results from a reliable method of the desired property can be used to predict this property for other compounds that have not been measured or synthesized. The use of appropriate mixing rules can calculate the heat of combustion of transportation fuels, such as naphtha, kerosene, and diesel from the heat of combustion of their constituent compounds when their compositions are known. This type of calculations is also possible for the use of surrogate fuels that simulate the components of the studied fuel [182, 183]. Thus, the knowledge of $\Delta H_c^o(\text{net})$ of the pure chemical compounds can determine the same property for undefined mixtures, such as petroleum fractions.

4.1 Experimental methods for determination of heats of combustion

Measurement of $\Delta H_c^o(\text{net})$ is complicated by the existence of several recognized ASTM standard test methods. Available ASTM methods differ on the basis of the characteristics of the studied liquid. For example, ASTM D240-09 [8] can determine the heat of combustion of liquid hydrocarbon fuels and polymers. This test method covers the determination of the heat of combustion of liquid hydrocarbon fuels ranging in volatility from that of light distillates to that of residual fuels. $\Delta H_c^o(\text{net})$ is determined by burning a weighed sample in an oxygen bomb calorimeter under controlled conditions. The heat of combustion is computed from the temperatures observed before, during, and after combustion while allowing for the appropriate thermochemical and heat transfer corrections. ASTM D4809-13 [184] is used as a more precise method. For aviation fuels, ASTM D3338/D3338M-09 [185], ASTM D1405/D1405M-08 [186], and ASTM D4529-01 [187] are used. ASTM D4868-00 [188] is used for diesel and burner fuels.

The gross heat of combustion is normally determined by the oxygen bomb calorimeter method (Figure 4.1), which includes the heat of vaporization given up when the newly formed water vapor produced by oxidation of hydrogen is condensed and cooled to the temperature of the bomb. For nearly all industrial applications, this water vapor escapes as steam in the flue gases. Thus, water vapor is not available for useful work. To compensate for this loss, the net heat of combustion can be calculated by subtracting the latent heat of vaporization from the gross value obtained from the calorimeter, but this requires knowledge of the hydrogen content of the

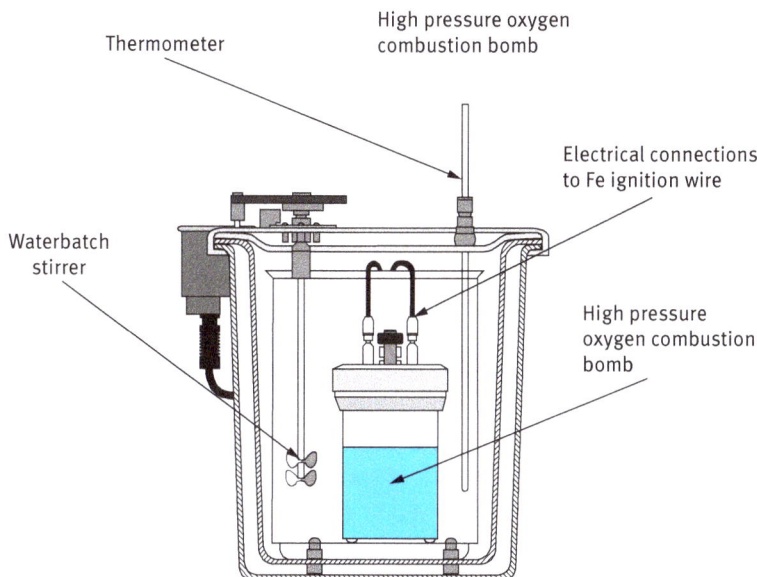

Figure 4.1: Schematic diagram of bomb calorimeter.

sample. Therefore, the net heat of combustion is calculated from the gross heat of combustion by assuming the formation of water is in a gaseous state. If the hydrogen content of a sample is known, the net heat of combustion at 298.15 K can be calculated as follows [189]:

$$\Delta H_{c}'(\text{net}) = \Delta H_{c}'(\text{gross}) - 212.2 \times H \tag{4.2}$$

where $\Delta H_{c}'(\text{net})$ is the net heat of combustion at constant pressure in kJ kg^{-1}; $\Delta H_{c}'(\text{gross})$ is the gross heat of combustion at constant volume in kJ kg^{-1}; H is the weight percent hydrogen in the sample. Due to the difficulty in accurate determination of the hydrogen content of the sample, and the fact that the hydrogen content of most fuels is fairly low, the gross heat of combustion is usually reported in preference to the net value for most applications.

4.2 Different approaches for prediction of the heats of combustion

Experimental determination of the heat of combustion of a new compound is too time-consuming because a complete reproducible combustion cannot be easily obtained. Predicting heat of combustion is valuable before expending resources. Various methods have been developed in the literature to estimate the heat of combustion [148, 180, 190–204]. Cardozo [190] proposed a group contribution method to estimate the heat of combustion for organic compounds using three simple correlations that depend on the state of the compound: solid, liquid, or gas. The necessary data are the total number of carbon atoms in the compound and the corrections for various structures and phases. The reported errors exceeded 12.5% for some compounds. This method can calculate the heat of combustion of complex organic compounds. Cardozo's model [190] relates the length of chain and heat of combustion based on 1,168 experimental data. It cannot be applied for polynitro-heteroarenes, acyclic and cyclic nitramines as well as energetic compounds with nitrate functional groups because group correction factors have not been specified for the compounds containing $N-NO_2$ and $O-NO_2$ functional groups and molecular fragments in polynitroheteroarenes. Seaton and Harrison [191] assumed some combustion products for pure or mixtures of compounds including 71 elements to estimate the heat of combustion. This approach is based on the Benson method [49] for predicting the heat of formation, which is considered to be too complex for manual calculations. Suitable group contribution and quantum mechanical methods can also be used for calculating the heat of formation or heat of sublimation, which can predict indirectly the heat of combustion [205–208]. Van Krevelen [192] used the heats of formation for the combustion products and reactants to calculate the heat of combustion. For polymeric reactants, the heat of combustion was estimated from the molar contributions of the chemical groups that constitute the monomer or repeat units. Kondo et al. [209] calculated the heats of formation for several flammable gases by the Gaussian-2 (G2) and/or G2MP2 method to obtain their heats of combustion and related constants for evaluating the combustion hazards. Quantum mechanical methods usually require special complex software, high-speed computers, and expert users. Hshieh [148] developed two empirical equations to predict the gross and the net heats of combustion of organosilicon compounds on the basis of the atomic contribution method. Hshieh and coworkers [194] have also introduced two other empirical equations for predicting the gross and net heats of combustion of organic polymers. Hshieh's methods [148, 194] have the advantage that only elemental compositions of organosilicon compounds and polymers are input parameters but they are only applicable to the organosilicon compounds and polymers. Diallo et al. [200] developed a model for calculating the heat of combustion of 53 ionic liquids using a multivariable linear regression technique.

QSPR models have also been used to predict heat of combustion of organic compounds. Gharagheizi [195] used genetic algorithm based GA-MLR to obtain four parameters multi-linear equation. This model is based on four complex molecular descriptors. Cao et al. [196] introduced another suitable QSPR for prediction of the standard net heat of combustion of organic compounds. This model is based on the atom-type electrotopological state (E-state) indices and ANN technique. Cao et al. [196] suggested a QSPR-ANN model to estimate the heat of combustion for pure organic compounds using atom-type electrotopological state indices with 49 structural groups as inputs. The model of Cao et al. [196] gives large AAEs for compounds containing fluorine and chlorine atoms. For compounds containing fluorine and chlorine, the model of Cao et al. [196] predicts the heat of combustion with an AAE of 615.98 and 299.65 kJ mol^{-1}, respectively [196]. Gharagheizi et al. [199] developed a complex method for calculating the heat of combustion of pure compounds from group contributions using a three-layered Feed-Forward Artificial Neural Network (FFANN) model with 142 intricate structural groups as inputs. Saldana et al. [210] created a consensus QSPR model for predicting the heat of combustion of 1,624 hydrocarbon-based compounds and 1,143 alcohols and esters. Saldana et al. [210] used various QSPR approaches to build models for predicting the heat of combustion ranging from methods leading to multi-linear models such as Genetic Function Approximation (GFA) and PLS to non-linear models, such as FFANN, General Regression Neural Networks (GRNN), SVM, and Graph Machines (GM). These models except the GM model use molecular descriptors and functional group count descriptors as inputs. Since all of the individual models have AAE of less than 2% except for the GRNN based model, which has an AAE of 4%. Saldana et al. [210] developed a consensus model by averaging the values computed with selected individual models to improve the generality and predictive power compared to individual predictive models. The robust consensus model can predict the heat of combustion with AAE = 0.7%. These QSPR models are based on complex descriptors, which require specific computer codes and expert users. Moreover, QSPR models may have some uncertainty and difficulty for complex molecular structures. A simple and reliable QSPR model has also been used for prediction of the heat of combustion of various energetic compounds including polynitro arene, polynitro heteroarene, acyclic and cyclic nitramine, nitrate ester, and nitroaliphatic compounds [197].

The methods of SGC are suitable methods for estimation of the heat of combustion of organic compounds. Sagadeev and coworkers [211, 212] have introduced some group contributions to calculate the heats of combustion For different classes of organic compounds. The method of Sagadeev and coauthors [211, 212] can be used only for certain types of nitroaromatic and nitroaliphatic compounds. Among different available methods for predicting the heat of combustion, several simple methods that provide reliable estimates and are easier to use are discussed in this chapter.

4.2.1 Predicting the standard net heat of combustion for pure hydrocarbons from their molecular structure

Albahri [202] introduced a group contribution method to predict the standard net heat of combustion of pure hydrocarbons from their molecular structures. He used a multivariable non-linear regression based on the least square method to arrive at a set of 32 atom-type structural groups. He represented the standard net heat of combustion for about 452 pure hydrocarbon substances. His method is very simple and requires no experimental data. This SGC can predict the standard net heat of combustion from the knowledge of the molecular structure alone with an AAE of 0.71%. It can also predict the standard net heat of combustion of hydrocarbon isomers as well.

$$\Delta H_c^o (\text{net}) = 154.608 + 1{,}177.8 \times \sum_i (\Delta H_c^o)_i - 0.00719 \times \left(\sum_i (\Delta H_c^o)_i \right)^3$$
$$+ 0.0053 \times \left(\sum_i (\Delta H_c^o)_i \right)^4$$

(4.3)

where $\sum_i (\Delta H_c^o)_i$ is the sum of the atom-type group contribution values for calculation of the standard net heat of combustion. The values of $(\Delta H_c^o)_i$ are given in Table 4.1.

Example 4.1: Calculate the standard net heat of combustion for p-diethyl benzene.

Answer: The molecular structure of p-diethyl benzene is given as:

It consists of two $-CH_3$ (Paraffin), two $>CH_2$ (Paraffin), four $=CH-$ (Aromatic), one $=C<$ (Aromatic), and one $=C<$ (Aromatic, $-p$). Thus, $\sum_i (\Delta H_c^o)_i$ is calculated as:

$$\sum_i (\Delta H_c^o)_i = 2 \times 0.5389 + 2 \times 0.5158 + 4 \times 0.4218 + 2 \times 0.3720 + 1 \times 0.3974 + 1 \times 0.3875$$
$$= 4.5815$$

The use of eq. (4.2) gives:

$$\Delta H_c^o (\text{net}) = 154.608 + 1{,}177.8 \times \sum_i (\Delta H_c^o)_i - 0.00719 \times \left(\sum_i (\Delta H_c^o)_i \right)^3$$
$$+ 0.0053 \times \left(\sum_i (\Delta H_c^o)_i \right)^4$$
$$= 154.608 + 1{,}177.8 \times 4.5815 - 0.00719 \times (4.5815)^3 + 0.0053 \times (4.5815)^4$$
$$= 5{,}552.34 \text{ kJ mol}^{-1}$$

The measured $\Delta H_c^o (\text{net})$ is 5,555.21 kJ mol^{-1} [43].

Example 4.2: Calculate the standard net heat of combustion for 6-n-propyl-[1,2,3,4-tetrahydronaphthalene].

Table 4.1: Atom-type structural groups and their contribution values for estimation of the standard net heat of combustion of pure hydrocarbons (kJ mol^{-1}) using eq. (4.3).

Serial No.	Group	$(\Delta H_c^o)_i$
1	–CH$_3$ (Paraffin)	0.5389
2	>CH$_2$ (Paraffin)	0.5158
3	>CH– (2–) (Paraffin)	0.4905
4	>CH– (3–) (Paraffin)	0.4963
5	>CH– (4–) (Paraffin)	0.4946
6	>CH– (5–) (Paraffin)	0.4963
7	>C< (2–) (Paraffin)	0.4700
8	>C< (3–) (Paraffin)	0.4748
9	=CH$_2$ (Olefin)	0.4923
10	=CH– (Olefin)	0.4713
11	=C< (2–) (Olefin)	0.4323
12	=C< (3–) (Olefin)	0.4566
13	=C= (Olefin)	0.4494
14	≡CH (Olefin)	0.4683
15	≡C– (Olefin)	0.4238
16	>CH$_2$ (Cyclic)	0.4948
17	>CH– (Cyclic)	0.4775
18	>CH– (–2)cis (Cyclic)	0.4831
19	>CH– (–2)trans (Cyclic)	0.4770
20	>CH– (–3)cis (Cyclic)	0.4753
21	>CH– (–3)trans (Cyclic)	0.4758
22	>CH– (–4)cis (Cyclic)	0.4657
23	>CH– (–4)trans (Cyclic)	0.4599
24	>C< (Cyclic)	0.4165
25	=CH– (Cyclic)	0.4371
26	=CH– (Aromatic)	0.4218
27	>CH$_2$ (Aromatic)	0.3651
28	=C< (Aromatic, fused)	0.3720
29	=C< (Aromatic)	0.3974
30	=C< (Aromatic, –o)	0.3898
31	=C< (Aromatic, –m)	0.3882
32	=C< (Aromatic, –p)	0.3875

Groups 29 to 32 are applied for non-fused. For non-cyclic compounds, the numbers 2–, 3–, 4–, and 5– show the carbon atom position along the hydrocarbon chain in the second, third, fourth, and fifth position, respectively, where they are calculated from the shortest distance from either end of the hydrocarbon chain. For cyclic compounds, the numbers 2–, 3–, and 4– refer to the second, third, and fourth position along the cyclic ring with respect to the reference group 17, respectively, where they are calculated from the shortest distance along the cyclic ring from either direction. For aromatics, the symbols o–, m–, and p– refer to the ortho, meta, and para positions on the aromatic ring.

Answer: The molecular structure of 6-n-propyl-[1,2,3,4-tetrahydronaphthalene] is given as:

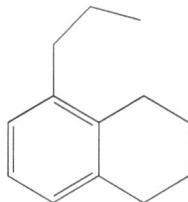

It consists of one $-CH_3$ (Paraffin), two $>CH_2$ (Paraffin), three $=CH-$ (Aromatic), four $>CH_2$ (Cyclic), one $=C<$ (Aromatic), one $=C<$ (Aromatic, $-o$), and one $=C<$ (Aromatic, $-p$). Thus, $\sum_i (\Delta H_c^o)_i$ is calculated as:

$$\sum_i (\Delta H_c^o)_i = 1 \times 0.5389 + 2 \times 0.5158 + 3 \times 0.4218 + 4 \times 0.4948 + 1 \times 0.3974 + 1 \times 0.3898$$
$$+ 1 \times 0.3898 + 1 \times 0.3882 = 5.99$$

The use of eq. (4.2) gives:

$$\Delta H_c^o(\text{net}) = 154.608 + 1{,}177.8 \times \sum_i (\Delta H_c^o)_i - 0.00719 \times \left(\sum_i (\Delta H_c^o)_i \right)^3$$
$$+ 0.0053 \times \left(\sum_i (\Delta H_c^o)_i \right)^4$$
$$= 154.608 + 1{,}177.8 \times 5.99 - 0.00719 \times (5.99)^3 + 0.0053 \times (5.99)^4 = 7{,}216 \text{ kJ mol}^{-1}$$

4.2.2 Prediction of the standard net heat of combustion from molecular structure

Albahri [204] developed a suitable QSPR method to predict $\Delta H_c^o(\text{net})$ of chemical compounds based only in their molecular structures. He used SGC method to estimate $\Delta H_c^o(\textit{net})$ through two models: a Multi-Variable Regression (MVR) based on least squares and an ANN. He applied the SGC method to probe the structural groups that have a significant contribution to the overall $\Delta H_c^o(\textit{net})$. He introduced 47 atom-type structural groups can represent the $\Delta H_c^o(\textit{net})$ for 586 pure substances. Among two approaches ANN and MVR, ANN was the more accurate. The method ANN can predict $\Delta H_c^o(\textit{net})$ with an average relative error of 0.89%. The results of the MVR model is less accurate, but it is also simple and practical and provides reliable estimates. The SGC method of Albahri [204] is very useful and convenient to assess the hazardous risks of chemicals. The MVR model uses a simple linear addition of the structural group contributions as:

$$\Delta H_c^o(\text{net}) = \sum_j (\Delta H_c^o)_j \qquad (4.4)$$

The values of $(\Delta H_c^o)_j$ are given in Table 4.2.

Example 4.3: Calculate the standard net heat of combustion for o-cresol by using eq. (4.3).

Answer: The molecular structure of p-diethyl benzene is given as follows:

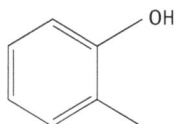

This compound consists of one $-CH_3$, four $=CH-$ (Ring), two $=C<$ (Ring), one $-OH$ (Phenol). Thus, the use of eq. (4.4) gives:

Table 4.2: The structural groups and their contribution values for estimation of the standard net heat of combustion of pure compounds (kJ mol^{-1}) using eq. (4.4).

Serial No.	Group	$(\Delta H_c^o)_j$
1	$-CH_3$	711.40
2	$>CH_2$	609.28
3	$>CH-$	503.00
4	$>C<$	398.87
5	$=CH_2$	667.53
6	$=CH-$	544.47
7	$=C<$	431.42
8	$=C=$	503.17
9	$\equiv CH$	621.61
10	$\equiv C-$	502.60
11	$-CH_2$ (Ring)	612.64
12	$>CH-$ (Ring)	519.26
13	$>C<$ (Ring)	425.69
14	$=CH-$ (Ring)	522.59
15	$=C<$ (Ring)	416.78
16	$-F$	−101.00
17	$-Cl$	−49.50
18	$-Br$	−51.00
19	$-OH$ (Alcohol)	−88.00
20	$-O-$ (Non-ring)	−106.00
21	$>C=O$ (Ketone)	230.28
22	$-CH=O$ (Aldehyde)	371.63
23	$-COOH$ (Acid)	19.76
24	$-COO-$ (Ester)	33.50

(continued)

Table 4.2: (continued)

Serial No.	Group	$(\Delta H_c^0)_j$
25	=O (Except as above)	0.1
26	$-NH_2$	257.91
27	>NH (Non-ring)	126.05
28	>N– (Non-ring)	71.89
29	–OH (Phenol)	–110.00
30	–O– (Ring)	–115.00
31	>NH (Ring)	60.81
32	–N= (Ring)	2.31
33	>N– (Ring)	244.67
34	–H	120.17
35	≡C	228.36
36	≡O	54.50
37	=S	278.58
38	>S	326.55
39	–SH	417.24
40		219.52
41	≡N	153.34
42	=S= (Non-ring)	90.31
43		0.02
44	–O–	4.40
45		220.57
46		–76.00
47		187.93

$$\Delta H_c^o \text{(net)} = \sum_j \left(\Delta H_c^o\right)_j = 1 \times 711.40 + 4 \times 522.59 + 2 \times 416.78 + 1 \times (-110.00)$$
$$= 3{,}525.3 \text{ kJ mol}^{-1}$$

The measured ΔH_c^o(net) is 3,517.45 kJ mol^{-1} [43].

Example 4.4: Calculate the standard net heat of combustion for n-eicocylcyclohexane by using eq. (4.4).

Answer: The molecular structure of n-eicocylcyclohexane is given as follows:

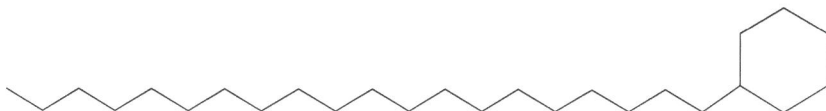

This compound consists of one –CH$_3$, 19 >CH$_2$, one >CH– (Ring), and five –CH$_2$ (Ring). Thus, the use of eq. (4.4) gives:

$$\Delta H_c^o \text{(net)} = \sum_j \left(\Delta H_c^o\right)_j = 1 \times 711.40 + 19 \times 609.28 + 1 \times 519.26 + 5 \times 612.64$$
$$= 15{,}870 \text{ kJ mol}^{-1}$$

The measured ΔH_c^o(net) is 15,970.80 kJ mol^{-1} [43].

4.2.3 A new method for predicting the gross heat of combustion of polynitro arene, polynitro heteroarene, acyclic and cyclic nitramine, nitrate ester and nitroaliphatic compounds

A reliable method has been presented for estimating the gross heats of combustion of important classes of energetic compounds including polynitro arene, polynitro hetero-arene, acyclic and cyclic nitramine, nitrate ester and nitroaliphatic compounds. It is based on elemental compositions as well as the presence of some specific polar groups and molecular fragments. It can be easily used for any complex organic compounds with at least one nitro, nitramine, or nitrate functional groups, where the predictions of their heats of combustion by the available methods are inaccurate or difficult. This model gives reliable predictions of heats of combustion with respect to QSPR and SGC methods where they can be applied. The model has the following form [197]:

$$\Delta H_c^o \text{(gross)} = 418.3 n_C + 109.7 n_H + 107.5 n_N + 67.88 n_O - 237.2 \text{Dec(polar groups)} \\ + 89.42 \text{Inc (molecular fragments)} \tag{4.5}$$

where ΔH_c^o(gross) is in kJ mol^{-1}; *Dec(polar groups)* and *Inc(molecular fragments)* show the presence of some specific polar groups and molecular fragments in decreasing and increasing the value of the heat of combustion on the basis of elemental composition, respectively. It was indicated that the presence of some specific polar groups may affect the predicted values of the condensed phase heat of formation of different classes of energetic compounds because high intermolecular attractions [213– 227]. The existence of some molecular fragments may have opposite effects and the adjustment of the contribution of the elemental composition is necessary. The values of *Dec*(*polar groups*) and *Inc*(*molecular fragments*)can be specified according to the following situations:

a) *Dec*(*polar groups*): The attachment of some specific polar groups such as –COOH to nitroaromatic compounds can decrease the heat of formation of organic compounds containing energetic functional groups. For the existence of one of three polar functional groups –OH, –COOH and –N–C(=O)- or more than one – NH$_x$ attached to nitroaromatic ring such as 1,3-diamino-2,4,6-trinitrobenzene (DATB), the value of *Dec*(*polar groups*) equals 1.0. This situation can enhance molecular interactions.

b) *Inc*(*molecular fragments*): The attachment of –N=N- to nitroaromatic compound such as 4,4′-dinitroazofurazan can increase the condensed phase heat of formation. The value of *Inc*(*molecular fragments*) in this situation or the existence of molecular fragment $\overset{}{\underset{+}{\diagdown}N{-}O}$ in polynitro heteroarenes, e.g. benzotrifuroxane, is equal to 2.0. The value of *Inc*(*molecular fragments*) is also 0.8 for mononitro benzene without the presence of polar groups of part (a).

For 121 energetic compounds, the values of R^2 and rms deviation mainly reflect the goodness of fit of the models, which are 0.999 and 60 kJ mol^{-1}, respectively, for eq. (4.5). If the above conditions to assign different values are not met, the values of *Dec*(*polar groups*) and *Inc*(*molecular fragments*) are equal to zero.

Example 4.5: Calculate the standard gross heat of combustion for the following compounds by using eq. (4.5).
(a) 7-oxabicyclo[4.1.0]heptane-2,3,4,5-tetrayl tetranitrate
(b) 3,5-dinitro-N2,N6-bis(2,4,6-trinitrophenyl)pyridine-2,6-diamine
(c) (E)-1-(4-nitro-1,2,5-oxadiazol-3-yl)-2-(4-((E)-2-(4-nitro-1,2,5-oxadiazol-3-yl)-1-oxi-dodiazen-1-yl)-1,2,5-oxadiazol-3-yl)diazene oxide

Answer: (a) The molecular structure of 7-oxabicyclo[4.1.0]heptane-2,3,4,5-tetrayl tetranitrate is given as follows:

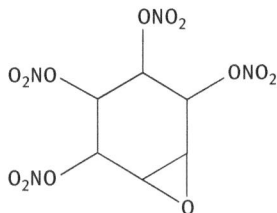

Since $n_C = 6$, $n_H = 6$, $n_N = 4$, $n_O = 13$, $Dec(polar\ groups) = 0$ and $Inc(molecular\ fragments) = 0$ for this compound, the use of eq. (4.5) gives:

$$\Delta H_c^o(gross) = 418.3 n_C + 109.7 n_H + 107.5 n_N + 67.88 n_O - 237.2\,Dec\,(polar\ groups)$$
$$+ 89.42\,Inc(molecular\ fragments)$$
$$= 418.3 \times 6 + 109.7 \times 6 + 107.5 \times 4 + 67.88 \times 13 - 237.2 \times 0 + 89.42 \times 0$$
$$= 2{,}716\ kJ\ mol^{-1}$$

The measured $\Delta H_c^o(gross)$ is 2,774 kJ mol^{-1} [102].
(b) The molecular structure of 3,5-dinitro-N2,N6-bis(2,4,6-trinitrophenyl)pyridine-2,6-diamine is given as follows:

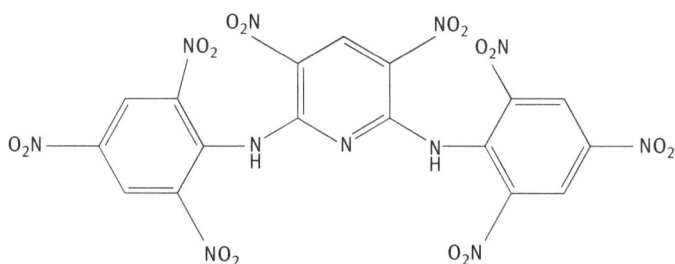

Since $n_C = 17$, $n_H = 7$, $n_N = 11$, $n_O = 16$, $Dec(polar\ groups) = 1.0$ and $Inc(molecular\ fragments)=0$ for this compound, the use of eq. (4.5) gives:

$$\Delta H_c^o(gross) = 418.3 n_C + 109.7 n_H + 107.5 n_N + 67.88 n_O - 237.2\,Dec(polar\ groups)$$
$$+ 89.42\,Inc\,(molecular\ fragments)$$
$$= 418.3 \times 17 + 109.7 \times 7 + 107.5 \times 11 + 67.88 \times 16 - 237.2 \times 1 + 89.42 \times 0$$
$$= 7{,}739\ kJ\ mol^{-1}$$

The measured $\Delta H_c^o(gross)$ is 7,769 kJ mol^{-1} [228].
(c) The molecular structure of (E)-1-(4-nitro-1,2,5-oxadiazol-3-yl)-2-(4-((E)-2-(4-nitro-1,2,5-oxadiazol-3-yl)-1-oxidodiazen-1-yl)-1,2,5-oxadiazol-3-yl)diazene oxide is given as follows:

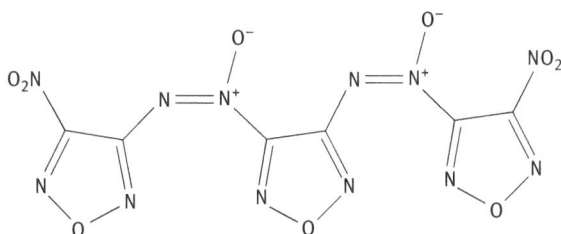

Since $n_C = 6$, $n_H = 0$, $n_N = 12$, $n_O = 9$, $Dec(polar\,groups) = 0$ and $Inc(molecular\,fragments) = 2.0$ for this compound, the use of eq. (4.5) gives:

$$\Delta H_c^o(\text{gross}) = 418.3 n_C + 109.7 n_H + 107.5 n_N + 67.88 n_O - 237.2\,Dec(\text{polar groups})$$
$$+ 89.42\,Inc(\text{molecular fragments})$$
$$= 418.3 \times 6 + 109.7 \times 0 + 107.5 \times 12 + 67.88 \times 9 - 237.2 \times 0 + 89.42 \times 2$$
$$= 3{,}368\,\text{kJ mol}^{-1}$$

The measured $\Delta H_c^o(\text{gross})$ is 3,451 kJ mol^{-1} [102].

Example 4.6: Calculate the standard gross and net heat of combustion for 2,4,6-trinitrotoluene with the following compounds by using eq. (4.5).

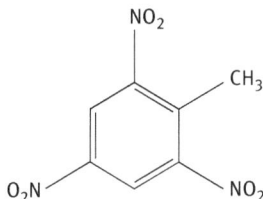

Answer: Since $n_C = 7$, $n_H = 5$, $n_N = 3$, $n_O = 6$, $Dec(polar\,groups) = 0$ and $Inc(molecular\,fragments) = 0$ for this compound, the use of eq. (4.5) gives:

$$\Delta H_c^o(\text{gross}) = 418.3 n_C + 109.7 n_H + 107.5 n_N + 67.88 n_O - 237.2\,Dec(\text{polar groups})$$
$$+ 89.42\,Inc(\text{molecular fragments})$$
$$= 418.3 \times 7 + 109.7 \times 5 + 107.5 \times 3 + 67.88 \times 6 - 237.2 \times 0 + 89.42 \times 0$$
$$= 3{,}392\,\text{kJ mol}^{-1}$$

Two reported experimental data for $\Delta H_c^o(\text{gross})$ are 3,389 and 3,395 kJ mol^{-1} [102]. Equation (4.2) can be rewritten as:

$$\Delta H_c'(\text{net}) = \Delta H_c'(\text{gross}) - 212.2\,\text{kJ kg}^{-1} \times \text{H}$$

$$\Delta H_c'(\text{net}) \times \frac{\text{Mw}}{1{,}000} = \Delta H_c'(\text{gross}) \times \frac{\text{Mw}}{1{,}000} - 212.2\,\text{kJ kg}^{-1} \times H \times \frac{\text{Mw}}{1{,}000}$$

$$\Delta H_c^o(\text{net}) = \Delta H_c^o(\text{gross}) - 212.2\,\text{kJ kg}^{-1} \times H \times \frac{\text{Mw}}{1{,}000}$$

$$\Delta H_c^o(\text{net}) = \Delta H_c^o(\text{gross}) - 21.43\,\text{kJ mol}^{-1} \times n_H$$

where *Mw* is the molecular weight of the desired compound. Thus, the net heat of combustion of 2,4,6-trinitrotoluene is calculated as:

$$\Delta H_c^o(\text{net}) = \Delta H_c^o(\text{gross}) - 21.43 \text{ kJ mol}^{-1} \times n_H$$
$$= 3{,}392 - 21.43 \times 5 = 3{,}285 \text{ kJ mol}^{-1}$$

The measured $\Delta H_c^o(\text{gross})$ is 3,295.9 kJ mol^{-1} [102].

4.3 Summary

This chapter introduces different approaches for the simple prediction of gross and net heats of combustion of pure hydrocarbons as well as some classes of pure compounds. Most of available predictive models usually require complex molecular descriptors, computer codes, and the expert users. Two SGC methods have been introduced for prediction of the heat of combustion of hydrocarbons and pure compounds. Equation (4.2) and corresponding Table 4.1 provide a simple path for the accurate estimation of the net heat of combustion of pure hydrocarbons. Equation (4.3) and corresponding Table 4.2 extend the previous method not only for pure hydrocarbons but also for the other classes of pure compounds such as alcohols and carboxylic acids. Equation (4.4) also provides a reliable simple relationship for the simple prediction of gross and net heats of combustion of important classes of CHNO energetic compounds including polynitro arene, polynitro heteroarene, acyclic and cyclic nitramine, nitrate ester, and nitroaliphatic compounds.

5 Polymer Flammability

The burning rate, ignitability, flammability, and heat release rate of polymers are not their intrinsic properties. These properties are extrinsic quantities resulting from the reaction of a macroscopic polymer sample due to severe thermal exposure. Thus, no single material property has been correlated with fire performance, nor does any test measure fire performance unambiguously for polymer flammability. Flammability of polymeric materials is an important area of research and development. Assessment of the combustion properties of polymeric materials can be done based on some important parameters like total heat release, specific heat release rate, and heat release capacity [229]. Microscale combustion calorimetry (MCC) can be used to determine these parameters. It is a small-scale flammability testing technique to screen polymer flammability prior to scale-up. Thermogravimetric analysis (TGA) and derivative TGA (DTGA) data of polymers can confirm the measured data of the MCC [230]. Specific heat release rate is the molecular-level fire response of a burning polymer, which can be measured by analyzing the oxygen consumed through the complete combustion of the pyrolysis gases during a linear heating program. Diving specific heat release rate to the rate of increase in temperature of a sample during a test can determine its heat release capacity. Heat release capacity is a combination of thermal stability and combustion properties. It can be calculated by the following equation [231–233]:

$$\text{HRC} = \frac{Q_c^\circ (1-\mu) E_a}{eRT_p^{\,2}} \tag{5.1}$$

where HRC is the heat release capacity; Q_c is the heat of complete combustion of the pyrolysis gases; μ is the weight fraction of the solid residue after pyrolysis or burning; E_a is the global activation energy for the single-step mass loss process of pyrolysis; T_p is the temperature at the peak mass-loss rate in a linear heating program at a constant rate; e is the natural number; and R is the gas constant. The heat of combustion per unit mass of original polymer, h_c ($J\,g^{-1}$), is also defined as the heat of combustion per unit mass of fuel gases multiplied by the mass fraction of volatile fuel generated during pyrolysis, which can be written as [229]:

$$h_c = (1-\mu)Q_c. \tag{5.2}$$

Combustion of a wide range of organic compounds and common polymers produces 13.1 ± 0.7 kJ of heat per gram of diatomic oxygen consumed. Thus, this value is independent of the chemical composition of the organic material [193, 234, 235]. The gases produced during polymer decomposition are usually unknown. Thus, they do not burn to completion in real fires. Oxygen consumption is a means of measuring heat release without detailed knowledge of the fuel species.

https://doi.org/10.1515/9783110572223-005

5.1 Experimental method based on pyrolysis combustion flow calorimetry

The method of pyrolysis combustion flow calorimetry (PCFC) can be used to study flammability of polymers [236, 237], which is shown in Figure 5.1. Some authors have used this method to report flammability data of polymers. For example, Walters and Lyon [236] and Zhang et al. [237] reported flammability data for some polymers. The procedures used by these authors are discussed here. Walters and Lyon [236] measured flammability characteristics of 86 polymers by a pyrolysis probe (Pyroprobe 2000, CDS Analytical, Oxford, PA) to thermally decompose milligram-sized samples in flowing nitrogen at a controlled heating rate. They heated samples (1.0 ± 0.1 mg) typically at 4.3 °C/s, starting from a temperature which is several degrees below the onset degradation temperature of the polymer to a maximum temperature of 1,203 K. The final temperature (1,203 K) was selected to ensure complete thermal degradation of organic polymers. Flowing nitrogen swept the volatile decomposition products from the pyrolysis chamber. Walters and Lyon [236] added oxygen for 60 s to obtain a nominal composition of 4:1 N_2/O_2 in a 1,173 K furnace. The combustion products containing carbon dioxide, water, and possibly acid gases were usually removed from the gas stream by Ascarite and Drierite scrubbers. Mass flowmeter and zirconia oxygen analyzer (Panametrics model 350, Waltham, MA) were used to measure the

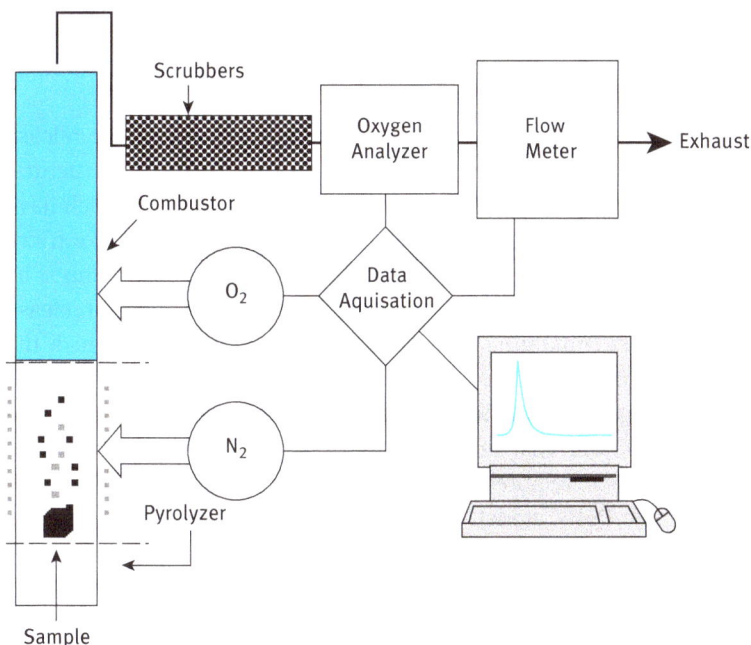

Figure 5.1: Schematic diagram of the pyrolysis combustion flow calorimetry (PCFC).

mass flow rate and oxygen consumption of the scrubbed combustion stream, respectively. Zhang et al. [237] also used the similar experimental procedure as Walters and Lyon [236] to study flammability of polymers. They pyrolyzed the samples of 1.0 ± 0.1 mg in a commercial device (CDS Pyroprobe 2000) to 1,203 K at 4.3 °C/s in the presence of N_2. The consumption rate of O_2 was measured continuously where the volatiles were swept away by an N_2/O_2 and were completely combusted at 1,173 K. They reported the heat release results for each sample as the average of five measurements. Pyrolysis gas chromatography-mass spectrometry (GC-MS) was applied to analyze the composition of the volatiles. They pyrolyzed samples of 0.2–0.3 mg to 930 °C at 4.3 °C/s in helium. The volatiles from pyrolysis were then separated by a Hewlett-Packard 5890 Series II gas chromatograph. The volatiles were also analyzed by a Hewlett-Packard 5972 Series mass spectrometer. Zhang et al. [237] analyzed the total heat release for several polymers such as polyethylene (PE) and polystyrene (PS) where their reported data were close to those presented by Walters and Lyon [236]. Appendix D gives flammability data including heat release capacity ($J\ g^{-1}\ K^{-1}$), total heat release (kJ g^{-1}), and char yield (%) for some polymers. Experimental data given in Appendix D were taken from different sources, i.e., references [236] and [237–240]. About 75% of the data in Appendix D are taken from Walters and Lyon [236] as well as Zhang et al. [237]. The remaining data are also consistent with those determined by PCFC method.

5.2 Different approaches for prediction of flammability parameters

For prediction of total heat release, specific heat release rate, and heat release capacity, SGC and QSPR methodology based on complex descriptors or structural parameters of repeat units of polymers are the two different approaches which have been developed recently [236, 241–243]. The SGC method uses an additive contribution from a variety of functional groups and molecular fragments. It cannot be applied for prediction of total heat release, specific heat release rate, and heat release capacity of those polymers containing a particular functional group for which the value of group contribution has not been defined. Parandekar et al. [243] used QSPR approaches based on complex or unfamiliar descriptors to predict total heat release, specific heat release rate, and heat release capacity using genetic function algorithms. This QSPR approach requires specific computer codes and expert users [243]. Simple and reliable QSPR models based on structural parameters have also been introduced to predict the values of total heat release, specific heat release rate, and heat release capacity of different polymers [244]. In this chapter, SGC methods and the simplest models based on structural moieties are introduced for prediction of flammability of different polymers with their repeat units, which are comprised of different chemical groups or moieties.

5.2.1 SGC method of Walters and Lyon for prediction of the heat release capacity

Walters and Lyon [236] introduced 42 different empirical molar group contributions to the heat release capacity. They compared the measured and calculated values of heat release capacity for over 80 polymers where AAE = 15%. Table 5.1 shows the structural groups and their molar contributions to the heat release capacity. For calculation of the heat-release capacity, the following equation is used:

$$\eta_c = \frac{\sum_i b_i \psi_i}{\sum_i b_i MM_i} \tag{5.3}$$

where η_c is the heat release capacity on a mass basis J g^{-1} K^{-1}; ψ_i is the molar heat-release capacity in kJ mol^{-1} K^{-1}; MM_i is the molar mass of the repeat unit in g mol^{-1}; and b_i is the number of the specified group of the repeat unit.

Table 5.1: Group contributions for estimating the heat release capacity.

Group	$\psi_i/$(kJ mol^{-1} K^{-1})	Group	$\psi_i/$(kJ mol^{-1} K^{-1})
	118	—O—	−11.6
	77.0	—N=P— (with O above and below P)	−13.8
	69.5	—NH$_2$	−13.9
	30.6	—CF$_3$	−14.8
	29.5	—CN	−17.6

(continued)

Table 5.1: (continued)

Group	$\psi_i/(\text{kJ mol}^{-1}\text{K}^{-1})$	Group	$\psi_i/(\text{kJ mol}^{-1}\text{K}^{-1})$
	28.8		−18.9
	28.3		−19.2
—CH	26.6	—OH	−19.8
—CH$_3$	22.5	—Br	−22.0
	19.0		−22.0
	18.7		−23.2
	16.7		−25.5
	15.1	—Cl	−34.7
	9.7		−36.4
—H	8.1		Pendant: −39.5 Backbone: −13.7

Table 5.1: (continued)

Group	$\psi_i/(\text{kJ mol}^{-1}\,\text{K}^{-1})$	Group	$\psi_i/(\text{kJ mol}^{-1}\,\text{K}^{-1})$
$\diagdown$NH	7.6	$-$N$\diagup$	-43.0
$-\overset{H_2}{\underset{}{C}}-O-$	4.18	$-O-\overset{O}{\overset{\|}{C}}-O-$	-49.0
$\diagup$CF$_2$	1.8	$-$Si$-$	-53.5
Cl$-$C$=$C$-$Cl	0.1	triazine ring	-66.7
benzimidazole	-8.8	triazine (trioxy) ring	-74.5
$-$S$-$	-10.9	$-O-\overset{O}{\underset{O}{\overset{\|}{\underset{\|}{P}}}}-O-$	-76.7

Example 5.1: Calculate the heat release capacity for a diglycidyl ether of bisphenol A (BPA epoxy), which is cured by anionic ring-opening polymerization.

Answer: The BPA epoxy has the following molecular structure:

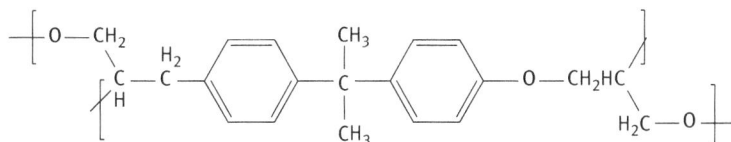

The contributions of various groups are given as follows:

Chemical group (i)	b_i	MM_i/(g/mol)	ψ_i/(kJ mol^{-1} K^{-1})	$b_i MM_i$/(g mol^{-1})	$b_i\psi_i$/(kJ mol^{-1} K^{-1})
$-\overset{\mid}{\underset{\mid}{C}}-$	1	12	28.3	12	28.3
$-\overset{\mid}{CH}-$	2	13	26.6	26	53.2
$>CH_2$	4	14	16.7	56	66.8
$-CH_3$	2	15	22.5	30	45.0
(benzene ring)	2	76	28.8	152	57.6
$-O-$	4	16	−11.6	64	46.4
Total				340	204.5

The use of eq. (5.3) gives:

$$\eta_c = \frac{\sum_i b_i\psi_i}{\sum_i b_i MM_i} = \frac{204.5\,\text{kJ mol}^{-1}\text{K}^{-1}}{340\,\text{g mol}^{-1}} = 601\,\text{J g}^{-1}\text{K}^{-1}.$$

The measured value of η_c is 657 J g^{-1} K^{-1} [236].

5.2.2 SGC method of Lyon et al. for prediction of total heat release (heat of combustion), char yield, and heat release capacity

Lyon et al. [242] introduced empirical molar group contributions for prediction of total heat release (heat of combustion), char yield, and heat release capacity of polymers. The following equations, beside eq. (5.3), are used to estimate these properties:

$$h_c = \frac{\sum_i b_i\Omega_i}{\sum_i b_i MM_i} \tag{5.4}$$

Table 5.2: Group contributions for estimating ψ_i, Ω_i, and X_i.

Chemical group[a]	Type[b]	Molar group contributions		
		$\Omega_i/(MJ\ mol^{-1})$	$X_i/(g\ mol^{-1})$	$\psi_i/(kJ\ mol^{-1}\ K^{-1})$
CH₃ group (C with two CH₃)	B	2.63	0	85.5
—CH₃	S	0.50	0	18.5
—CH₂—	B	0.51	0	14.4
naphthalene group	S	4.84	0	108.6
—O—	B	0.12	0	11.1
dimethyl benzene group	B	2.19	21	43.2
pyridine group	S	2.16	0	39.6
—CH—	S	0.67	0	5.7
—NH—	B	0.72	0	6.1
benzene R group	S,B	1.69	21	24.7
naphthalene group	B	2.38	54	39.6
dimethyl benzene group	B	−0.69	76	23.0
—OH	S	−0.12	0	4.8

(continued)

Table 5.2: (continued)

Chemical group[a]	Type[b]	Molar group contributions		
		$\Omega_i/(\text{MJ mol}^{-1})$	$X_i/(\text{g mol}^{-1})$	$\psi_i/(\text{kJ mol}^{-1}\,\text{K}^{-1})$
—S—	B	0.78	9	7.9
—O—C(=O)—O—	B	0.22	−1	13.6
—CH$_2$O—	B	0.42	0	5.8
(para-phenylene)	B	1.08	35	13.1
(meta-phenylene)	B	1.08	35	13.1
(phenyl)	S	1.49	37	12.5
—C(=O)—O—	B	0.24	0	3.1
—CF$_2$—	B	0.15	2	2.5
(substituted benzene with R groups)	S, B	1.2	48	0.4
(benzimidazole)	B	0.79	87	−1.0
(bisoxazole)	B	0.19	128	−3.3
—CF$_3$	S	−0.96	40	−4.0
—NH$_2$	S	−0.04	1	−1.4
>N—	S, B	0.13	0	−1.7

Table 5.2: (continued)

Chemical group[a]	Type[b]	Molar group contributions		
		$\Omega_i/(\text{MJ mol}^{-1})$	$X_i/(\text{g mol}^{-1})$	$\psi_i/(\text{kJ mol}^{-1}\,\text{K}^{-1})$
(phthalimide group)	B	−0.04	100	−1.7
(pyromellitic diimide group)	B	0.25	1.29	−27.7
(benzoyl group)	S	0.27	96	−17.3
(sulfone group, −S(=O)₂−)	B	0.31	−1	−10.9
(carbonyl ester, −C(=O)−O−)	S	−0.07	3	−13.5
—Cl	S	−0.47	11	−11.8
(−CH(CCl₂)− group)	S	−1.26	79	−33.3
(−C(CH₃)₂− group)	S	1.00	−2	−5.3
(carbonyl, −C(=O)−)	B	−0.19	7	−12.6
—CN	S	−1.85	26	−36.6
(−Si−)	B	−0.81	28	−98.5

[a] R=Organic substituent.
[b] B=Backbone group, S=Side or pendant group.

$$\mu\,(\%) = \frac{\sum\limits_{i} b_i X_i}{\sum\limits_{i} b_i MM_i} \times 100 \qquad (5.5)$$

where h_c is in J g^{-1}; $\mu(\%)$ is the percentage of char yield or the mass fraction of char (g-char/g-sample); and Ω_i and X_i are the corresponding molar group contributions. Table 5.2 shows the structural groups and their molar contributions on ψ_i, Ω_i, and X_i.

Example 5.2: Calculate total heat release (heat of combustion), char yield, and heat release capacity of polyethylene terephthalate (PET).

Answer: The PET has the following molecular structure:

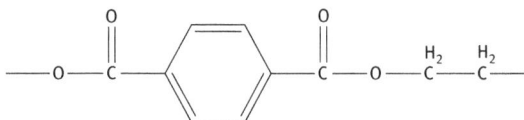

The contributions of various groups are given as follows:

Chemical group (*i*)	b_i	MM_i/(g mol^{-1})	Ω_i/(MJ mol^{-1})	X_i/(g mol^{-1})	ψ_i/(kJ mol^{-1} K^{-1})
	1	76	1.08	35	13.1
	2	44	0.24	0	3.1
	2	14	0.51	0	14.4

The use of eqs (5.3), (5.4), and (5.5) gives:

$$\eta_c = \frac{\sum\limits_{i} b_i \psi_i}{\sum\limits_{i} b_i M_i} = \frac{(1\times 13.1 + 2\times 3.1 + 2\times 14.4)\text{kJ mol}^{-1}\text{K}^{-1}}{(1\times 76 + 2\times 44 + 2\times 14)\text{g mol}^{-1}} = 251\,\text{J g}^{-1}\text{K}^{-1}$$

$$h_c = \frac{\sum\limits_{i} b_i \Omega_i}{\sum\limits_{i} b_i MM_i} = \frac{(1\times 1.08 + 2\times 0.24 + 2\times 0.51)\text{MJ mol}^{-1}}{(1\times 76 + 2\times 44 + 2\times 14)\text{g mol}^{-1}} = 0.0134\,\text{MJ g}^{-1} = 13.4\,\text{kJ g}^{-1}$$

$$\mu\ (\%) = \frac{\sum\limits_{i} b_i X_i}{\sum\limits_{i} b_i M_i} \times 100 = \frac{(1 \times 35 + 2 \times 0 + 2 \times 0)g\ mol^{-1}}{(1 \times 76 + 2 \times 44 + 2 \times 14)g\ mol^{-1}} \times 100 = 18\%.$$

The measured value of η_c, h_c, and $\mu\ (\%)$ are 350 J g^{-1} K^{-1}, 16 kJ g^{-1}, and 13%, respectively [242].

5.2.3 The simplest model for reliable prediction of total heat release (heat of combustion)

The simplest model has been introduced to predict the value of h_c of different polymers on the basis of their repeat units that may contain chemical groups or moieties. In contrast to the available QSPR methodology, the new method simply requires the only elemental composition of repeat units of the desired polymer without using complex molecular descriptors and computer codes where they need expert users. For the measured h_c of 122 polymers, the predicted results are compared with the calculated data of Lyon et al. [242]. The rms deviation, mean absolute deviation (MAD), and maximum observed error of the new model are 4.5, 3.4, and 15.3 kJ g^{-1}, respectively, which are lower than those predicted by the method used by Lyon et al. [242], i.e., 9.0, 5.8, and 35.7 kJ g^{-1}, respectively. The new correlation has the following form [245]:

$$h_c = 3.72 + \frac{307.90 n'_H - 323.77 n'_O - 18.040 n'_F - 543.32 n'_{Cl} - 56.252 n'_{Br}}{MW_{repeat\ unit}} \tag{5.6}$$

where h_c is in kJ g^{-1}; n'_H, n'_O, n'_F, n'_{Cl}, and n'_{Br} are the number of moles of hydrogen, oxygen, fluorine, chlorine, and bromine atoms per mole of repeat unit, respectively; $MW_{repeat\ unit}$ is the molecular weight of repeat unit in g mol^{-1}. For estimating the gross and net heats of combustion of important classes of energetic compounds including polynitro arene, polynitro heteroarene, acyclic and cyclic nitramine, nitrate ester, and nitroaliphatic compounds [197], elemental composition is also important. In contrast to what was expected, there is no contribution of carbon, nitrogen, phosphorous, silicon, and sulfur atoms in eq. (5.6). This abnormal situation is consistent with the fact that heats of combustion per gram of oxygen consumed for typical synthetic polymers are nearly constant, i.e., between 12.59 and 14.50 kJ per gram of O$_2$ [234]. Equation (5.6) considers the contribution of those atoms which have significant influence in adjusting heats of combustion per gram of oxygen consumed from the common value, i.e., 13.1 ± 0.7 kJ of heat per gram of diatomic oxygen consumed. It cannot be used for pure carbon or in a situation where $307.90 n'_H - 323.77 n'_O - 18.040 n'_F - 543.32 n'_{Cl} - 56.252 n'_{Br}$, because it may give large deviation, e.g., for graphite, where $n'_H = n'_O = n'_F = n'_{Cl} = n'_{Br} = 0$, which gives 3.72 kJ/g rather than 32.8 kJ/g [246].

Example 5.3: Calculate the value of h_c for poly(2-vinyl naphthalene).

Answer: Repeat unit has the following molecular structure:

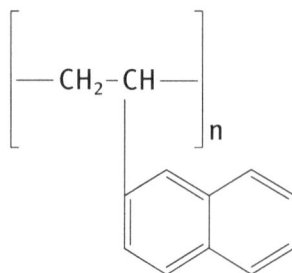

Thus, $n'_H=10$, $n'_O= n'_F= n'_{Cl}= n'_{Br}= 0$ and $MW_{repeat\ unit}= 154.22$ g mol^{-1}. Equation (5.6) gives the value of h_c as:

$$h_c = 3.72 + \frac{307.90(10)}{154.22} = 23.68 \text{ kJ g}^{-1}$$

The measured value of h_c is 16 kJ g^{-1} [236].

5.2.4 A simple model for reliable prediction of the specific heat release capacity of polymers

A simple model was introduced to predict the heat release capacity (η_c) of different polymers on the basis of their repeat units that may contain chemical groups or moieties, such as methyl, phenyl, carbonyl, ether, amide, and ester [244]. In contrast to available QSPR methods, it requires only the molecular structure of repeat unit of the desired polymer, without using complex molecular descriptors and computer codes where they need expert users. Model building has been constructed on the basis of the measured heat release capacity of 111 polymers and compared with the predicted results of the two group additivity methods. The rms deviation of the new model is 80 J g^{-1}K^{-1}, which is lower than those predicted by the two group additivity methods by Walters and Lyon [236] and Lyon et al. [242], i.e., 147 and 208 J g^{-1}K^{-1} corresponding to the application of group additivity methods for 110 and 101 polymers, respectively. For 11 newly synthesized polymers, it was shown that the rms value obtained by the new method was also lower than those obtained by group additivity methods. The correlation of η_c of the desired polymer can be obtained as follows [244]:

$$\eta_c = 97.00 + \frac{5,850n'_H - 17,532n'_N - 7495n'_O - 19,601n'_{Cl} - 83,828n'_{Si}}{MW_{repeat\ unit}}$$
$$+ 236.5\,HRC_{(CH_2)_n CXYZ,XArY} - 116.2 HRC_{YXArZ,Hal} \tag{5.7}$$

where η_c is in J g^{-1}K^{-1}; n'_N and n'_{Si} are the number of moles of nitrogen and silicon atoms per mole of repeat unit, respectively; $HRC_{(CH_2)_n CXYZ,XArY}$ and $HRC_{YXArZ,Hal}$ are the two correcting functions in J g^{-1}K^{-1} due to the presence of molecular fragments $-(CH_2)_{n\geq1}-C(X)(Y)-Z-$ or $-X-Ar-Y-$ and $Y-X-Ar-Z$ or halogens (fluorine) in repeat units. Aliphatic polymers, such as vinyl-based polymers, have higher energy gaps than aromatic or unsaturated polymers containing nitrogen and oxygen. Thus, the contribution of molecular fragment $-(CH_2)_{n \geq 1} - C(X)(Y) - Z-$ in these polymers can increase the value of $HRC_{(CH_2)_n CXYZ,XArY}$ significantly. As seen in eq. (5.7), the coefficients of all the atoms are negative except the coefficient of n'_H, which indicates that an increase in the values of n'_N, n'_O, n'_{Cl}, $and\ n'_{Si}$ in a desired polymer can decrease the value of η_c. Since the coefficient of n'_H is smaller than the coefficients of n'_N, n'_O, n'_{Cl}, $and\ n'_{Si}$, its contribution in lowering η_c is minor. The coefficients of electronegative elements n'_N, n'_{Cl}, and n'_{Si} are much higher than the coefficient of n'_O, which indicates that an increase in the values of n'_N, n'_{Cl}, and n'_{Si} is more effective in reducing η_c than an increase in the value of n'_O. An increment in the saturation level of a newly designed heat-resistant polymer is an appropriate way for decreasing the value of η_c because it also reduces the value of n'_H. The addition of relatively small amounts of silicon compounds to various polymeric materials can improve their flame retardancy [247–249]. The coefficient of n'_{Si} in eq. (5.7) has the largest negative value with respect to the coefficients of the other atoms. Introducing a silicon element and its groups to the monomers of suitable polymers, such as epoxy, can also improve some properties of the epoxy resins, such as thermal stability, high resistance to thermal oxidation, low surface energy and low toxicity [250, 251]. For phosphorus-containing compounds or resins, they are considered as effective flame retardants for epoxy resins. The presence of $-O-P$ can decrease the value of η_c with the contribution of $HRC_{YXArZ,Hal}$. The effects of halogens in decreasing the value of η_c appear in two terms: (a) n'_{Cl} for chlorine as additive term; and (b) $HRC_{YXArZ,Hal}$ for fluorine as non-additive contribution. For those polymers, such as biphenol phthalonitrile, where $HRC_{(CH_2)_n CXYZ,XArY}$ and $HRC_{YXArZ,Hal}$ have no contribution, and

$$\frac{5,850n'_H - 17,532n'_N - 7495n'_O - 19601n'_{Cl} - 83,828n'_{Si}}{MW_{repeat\ unit}} < 97,$$ the value of η_c

is 20 J g^{-1}K^{-1}. Table 5.3 gives the values of two correcting functions $HRC_{(CH_2)_n CXYZ,XArY}$ and $HRC_{YXArZ,Hal}$.

Example 5.4: Calculate the value of η_c for poly(2-vinyl naphthalene).

Answer: Repeating unit has the following molecular structure:

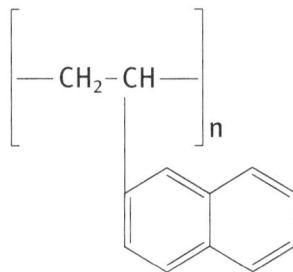

Table 5.3: The values of two correcting functions $HRC_{(CH_2)_n CXYZ,\, XArY}$ and $HRC_{YXArZ,\, Hal}$.

Predicting $HRC_{(CH_2)_n CXYZ,\, XArY}$

Molecular fragment in repeat unit	X	Y	Z	$HRC_{(CH_2)_n CXYZ,\, XArY}$	Example
$-(CH_2)_{n\geq1}-C(X)(Y)-Z-$	−H	−H	−CH₂−	3.3	Polyethylene (PE)
		−R		2.9	Polypropylene (PP)
		Carbocyclic aromatic without substituent		1.5	Polystyrene (PS)
		Pyridine		1.1	Poly(2-vinyl pyridene)
		−OH		0.4	Poly(vinyl alcohol) (99%; PVOH)
		Cyclic ether		1.1	Poly(vinyl butyral)
		−CN		0.8	Poly(acrylonitrile butadiene styrene) (ABS)
		−H	$-OCH_2^-$	0.9	Poly(ethylene oxide)
			$-C(O)\text{-}NH-$ or $-C(O)O\text{-}$	0.2	Polycaprolactam
			$-OC(O)Ar\text{-}O-$	0.7	Poly(ethylene terephthalate) (PET)
			$-NH-C(O)\text{-}(CH_2)_m-C(O)-$ where m≤8	1.6	Poly(hexamethylene adipamide)

Molecular moiety in repeat unit	X	Y	Z	HRC	Example
$-R$	$-R$		$-CH_2-$	0.4	Polyisobutylene
	$-COOH$			0.6	Poly(methacrylic acid)
	$-COOCH_3$			0.3	Poly(methyl methacrylate) (PMMA)
	Carbocyclic aromatic without substituent			0.6	Poly(α-methyl styrene)
$-X-Ar-Y-$	$-C(=CCl_2)-$	$-OC(O)-$	—	1.1	BPA Polyarylate
	$-CRR'-$	$-O-CH_2-$ or $-O-C=N$	—	0.6	EBPA/DDM

Estimating $HRC_{YXArZ, Hal}$

Molecular moiety in repeat unit	X	Y	Z	$HRC_{YXArZ, Hal}$	Example
$Y-X-Ar-Z$	$-CO-$	Ar	—	2.0	Poly(benzoyl 1,4phenylene)
	$-O-$ or the presence simultaneously of both $-O-$ and $-CO-$ (or $-SO2-$ or $-SO-$) without $-Ar-C(CH_3)_2-Ar-$		—	1.0	Poly(ether ketone) (PEK)
	$-O-P$	Ar or R		1.0	Polyphosphazene
		>CHCO	$-OC(O)-$	1.3	Chalcon II
$-CF_3$ or more than one $-CF_2$	—	—	—	0.7	Poly(tetrafluoro ethylene) (PTFE)

Thus, the values of elemental composition and $MW_{repeat\ unit}$ are $n'_H = 10$, $n'_N = n'_O = n'_{Cl}$ $= n'_{Si} = 0$ and $MW_{repeat\ unit} = 154.22$ g mol^{-1}. The value of $HRC_{(CH_2)_n CXYZ,XArY}$ is 1.5 J g^{-1}K^{-1} because $n = 1$, $X = -H$, Y = carbocyclic aromatic without substituent and $Z = -CH_2-$ for $-(CH_2)_{n \geq 1} - C(X)(Y) - Z -$. Since there is no contribution of $HRC_{YXArZ,Hal}$, the value of $HRC_{YXArZ,Hal} = 0$. Thus, eq. (5.7) gives the value of η_c as:

$$\eta_c = 97.00 + \frac{5850(10)}{154.22} + 236.5(1.5) = 831 \text{ J g}^{-1}\text{K}^{-1}$$

The measured value of η_c is 350 J g^{-1} K^{-1} [236].

5.2.5 A simple method for the reliable prediction of char yield of polymers

A simple model was introduced to predict the char yield of different polymers. It uses some molecular fragments besides elemental composition. In contrast to the QSPR methodology, the use of complex molecular descriptors, computer codes, and expert users is not required. The rms deviation of the model of 111 polymers was 7.2, which was lower than that predicted by the group additivity method, i.e., 16.3. The model has also been tested for 11 newly synthesized polymers where the rms values were lower than those obtained by the group additivity method. This model has the following form [252]:

$$\mu(\%) = 9.40 + 2.584n_C - 2.457n_H + 7.800n_N - 3.239n_F + 6.241n_{Cl} + 48.42n_{Si} \\ + 35.93\mu^+ - 33.21\mu^- \tag{5.8}$$

where n_C is the number of mole of carbon atoms per mole of repeat unit; μ^+ and μ^- are the two correcting functions where the char yield can be increased or decreased based

Table 5.4: The values of two correcting functions μ^+ and μ^-

Molecular fragment in repeat unit	X	Y	μ^+	μ^-
⬡—X	Cl_Cl C	–	0.3	0
	S, (N,N fused ring O,O)		0.7	0
	CF₃ C CF₃	–		
⬡—X—⬡—Y	O	CO or SO		

Table 5.4: (continued)

Molecular fragment in repeat unit	X	Y	μ^+	μ^-
[phosphate fragment $O{=}P(O)(O)$]	–	–	0.5	0
[phenyl–X–phenyl]	CO (without further functional groups), CH=N	–	1.0	0
$N{-}\overset{O}{\overset{\|}{C}}{-}(CH_2)n{-}\overset{O}{\overset{\|}{C}}$, $n \le 4$	–	–		
[pyridine ring]	–	–	0	0.7
[diphenyl phosphine oxide fragment]	–	–	0	1.0
[phenyl–X–phenyl]	$\overset{O}{\overset{\|}{C}}{-}\overset{H}{N}{-}\!\!\!-\!\!\!\overset{H}{N}{-}\overset{O}{\overset{\|}{C}}$	–		
[naphthalene fragment $O{=}C$ ... O]	–	–		

on the contribution of elemental composition. For negative values of $\mu(\%)$, there is no char yield, i.e., $\mu(\%) = 0.0$. The new method can predict char yield of the polymers containing carbon, hydrogen, nitrogen, fluorine, chlorine, silicon, oxygen, sulfur, phosphorous, and bromine atoms. For the presence of other atoms, the predicted results of eq. (5.8) may be large. Table 5.4 gives the values of two correcting functions μ^+ and μ^-.

Example 5.5: Calculate the value of $\mu(\%)$ for poly(4-vinyl pyridine).

Answer: Repeating unit has the following molecular structure:

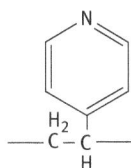

[structure: 4-vinyl pyridine repeat unit — pyridine ring with N at top, attached to $-CH_2-CH-$ backbone, labeled $H_2C{-}\underset{H}{C}$]

Thus, $n_C = 7$, $n_H = 7$, $n_N = 1$, and $n_F = n_{Cl} = n_{Si} = 0$ as well as $\mu^+ = 0$ and $\mu^- = 0.7$. The use of eq. (5.8) gives:

$$\mu(\%) = 9.40 + 2.584 n_C - 2.457 n_H + 7.800 n_N - 3.239 n_F + 6.241 n_{Cl} + 48.42 n_{Si}$$

$$+ 35.93 \mu^+ - 33.21 \mu^-$$

$$= 9.40 + 2.584\,(7) - 2.457\,(7) + 7.800\,(1) - 3.239\,(0)$$

$$+ 6.241\,(0) + 48.42\,(0) + 35.93\,(0) - 33.21\,(0.7)$$

$$= -5.2 \cong 0$$

The measured value of μ (%) is 13% [242].

5.3 Summary

Two different approaches – SGC and QSPR methods – were demonstrated in this chapter for prediction of heat release capacity (J g^{-1} K^{-1}), total heat release (kJ g^{-1}), and char yield (%). The reliability of SGC is less than QSPR methods based on molecular fragments. The contribution of n_H, n_N, n_O, n_{Cl}, and n_{Si} divided by MW$_{repeat\ unit}$ as well as two correcting functions HRC$_{(CH_2)_nCXYZ,XArY}$ and HRC$_{YXArZ,Hal}$ are needed for the estimation of heat release capacity. For reliable QSPR prediction of total heat release (heat of combustion) values, only the contributions of n_H, n_O, n_F, n_{Cl}, and n_{Br} divided by MW$_{repeat\ unit}$ are needed. Some structural fragments besides elemental composition have been included for more reliable prediction of the char yield.

Problems

(Hint: Experimental data for some problems are given in appendixes A to D)

Chapter 1

Use the following equations to calculate the flash point of the specified organic compounds:

1. Equation (1.3) for p-cresol, the NBP and $\Delta_{vpa}H$ are 475.133 K [43] and 47.22 kJ mol^{-1} [253]
2. Equation (1.4) for 1,10-decanediol, the NBP is 574 K [43]
3. Equations (1.5) and (1.6) for thiophen, the NBP is 357.3 K [43]
4. Equation (1.7) for 2,4-dimethylhexane, the NBP is 382.58 K [43]
5. Equation (1.9) for bis(dimethylamino)methylsilane
6. Equation (1.10) for n-nonylbenzene
7. Equation (1.11) for 2,3-dimethyl-3-ethylpentane
8. Equation (1.12) for n-octylbenzene
9. Equation (1.15) for 1-ethyl-1-methylcyclopentane
10. Equation (1.16) for deca-1,9-diene
11. Equation (1.17) for octan-1-amine
12. Equation (1.18) for 2-methoxyethanol
13. Equation (1.20) for 1-tetradecanol

Chapter 2

Use the following equations to calculate the autoignition temperature of the specified organic compounds:

1. Equation (2.1) for methyl vinyl ether
2. Equation (2.2) for tricosane
3. Equation (2.3) for 1-chloro-2,3-dinitrobenzene

Chapter 3

Use the following equations to calculate the upper and lower flammability limits of the specified organic compounds:

1. Equation (3.1) for prediction of the LFL (mol%) of furfuryl alcohol at 72°C
2. Equation (3.2) for prediction of the LFL (vol%) of trans-2-methyl-3-hexene
3. Equation (3.3) for prediction of the UFL (vol%) of n-hexadecylcyclohexane
4. Equation (3.4) for prediction of the UFL (vol%) of 1-butanol

https://doi.org/10.1515/9783110572223-006

Chapter 4

Use the following equations to calculate the heat of combustion of the specified organic compounds:
1. Equation (4.3) for prediction of the net heat of combustion of trans-3-methyl-3-heptene
2. Equation (4.4) for prediction of the net heat of combustion of n-propylamine
3. Equation (4.5) for prediction of the gross heat of combustion of N-(2,4,6-trinitrophenyl)-1H-1,2,4-triazol-3-amine
4. Equations (4.2) and (4.5) for prediction of the net heat of combustion of bis(4-nitrophenyl)amine

Chapter 5

Use the following equations to calculate the heat of combustion of the specified organic compounds:
1. Equation (5.3) and the method of Walters and Lyon [236] for prediction of the heat release capacity of poly(ethyl acrylate)
2. Equations (5.3), (5.4) and (5.5) and the method of Lyon et al. [242] for prediction of total heat release (heat of combustion), char yield and heat release capacity of poly(vinyl acetate) (PVAc)
3. Equations (5.6) for prediction of total heat release (heat of combustion) of poly (vinyl butyral)
4. Equation (5.7) for prediction of the heat release capacity of poly(tetrafluoro ethylene) (PTFE)
5. Equation (5.8) of the char yield of polylaurolactam

Answers to Problems

Chapter 1

1. 368 K
2. 432 K
3. 267 K and 266 K
4. 279 K
5. 268 K
6. 385 K
7. 292 K
8. 383 K
9. 279 K
10. 325 K
11. 337 K
12. 322 K
13. 397 K

Chapter 2

1. 586 K
2. 475 K
3. 680 K

Chapter 3

1. 1.73
2. 1.06
3. 11.81
4. 8.21

Chapter 4

1. 4975 kJ mol^{-1}
2. 2164 kJ mol^{-1}
3. 4003 kJ mol^{-1}
4. 5861 kJ mol^{-1}

https://doi.org/10.1515/9783110572223-007

Chapter 5

1. $484 \text{ J g}^{-1} \text{ K}^{-1}$
2. 10.9 kJ g^{-1}, 3.5% and $431 \text{ J g}^{-1} \text{ K}^{-1}$
3. 29.5 kJ g^{-1}
4. $16 \text{ J g}^{-1} \text{ K}^{-1}$
5. 0%

List of Symbols

AAE	Average absolute error
AIT	Autoignition temperature
AIT_{SPG}^{+}	Positive correcting function the presence of specific polar energetic groups in equation (2.3)
AIT_{SPG}^{-}	Negative correcting function the presence of specific polar energetic groups in equation (2.3)
ANN	Artificial neural networks
APE	Average percentage error
b_i	Number of the specified group of the repeat unit of polymer
CAMD	Computer-aided molecular design
CF	Correction factor in equation (1.18)
C_i	First-order group contribution of type i in equation (1.7)
C_{mix}	Stoichiometric concentration of the combustible components in the vapor phase
$Conc_i$	Stoichiometric concentration of the combustible component i
Dec(polar groups)	Presence of some specific polar groups for decreasing the value of heat of combustion in equation (4.5)
D_j	Second-order group contribution of type j in equation (1.7)
DP	Decreasing parameter based on structural parameters of amines given in equation (1.17)
DSP	Correcting function for decreasing flash point of hydrocarbons in equation (1.11)
DTG	Derivative thermogravimetric analysis
e	Natural number
E_a	Global activation energy for the single-step mass-loss process or pyrolysis
F_{SH}	Correcting function for branches of different classes of hydrocarbons in equation (2.2)
f_i	Group contribution for the ith contributed group in equations (1.9) and (2.1)
FFANN	Feed-Forward Artificial Neural Network
FP	Flash point
FP_i	Atom-type structural group contribution in equation (1.10)
$FP^{(+)}$	Contribution of structural parameters of unsaturated hydrocarbons for increasing the flash point on the basis of $(FP)_{core}$ in equation (1.14)

https://doi.org/10.1515/9783110572223-008

FP^+	Increasing flash point parameter given in equation (1.16)
$FP^{(-)}$	Contribution of structural parameters of unsaturated hydrocarbons for decreasing the flash point on the basis of $(FP)_{core}$ in equation (1.14)
FP^-	Decreasing flash point parameter given in equation (1.16)
$(FP)_{core}$	Linear function of the number of carbon and hydrogen atoms in equation (1.11)
$(FP)_{correcting}$	Correcting flash point function given in equation (1.14)
F_{SH}	Correcting function for size of different classes of hydrocarbons in equation (2.2)
GA-MLR	Genetic Algorithm-based Multivariate Linear Regression
GFA	Genetic Function Approximation
g_i	Structural group contribution of the molecule in equations (1.19) and (1.20)
GM	Graph Machines
GRNN	General Regression Neural Networks
GP	Genetic programming
H	Weight percent hydrogen in the sample
h_c	Heat of combustion (total heat release) per unit mass of original polymer
HBD	Hydrogen Bonding Donors
HRC	Heat release capacity
$HRC_{(CH_2)_n CXYZ,XArY}$	Correcting function of specific heat-release capacity for the presence of molecular fragment $-(CH_2)_{n \geq 1} - C(X)(Y) - Z -$ or $-X - Ar - Y -$ in repeat units of polymer
$HRC_{YXArZ,Hal}$	Correcting function of specific heat-release capacity for the presence of molecular fragments $Y - X - Ar - Z$ or halogens (fluorine) in repeat units of polymer
Inc(molecular fragments)	Presence of some specific molecular fragments for increasing the value of heat of combustion in equation (4.5)
IP	Increasing parameter based on structural parameters of amines given in equation (1.17)
ISP	Correcting function for increasing flash point of hydrocarbons in equation (1.11)
LCC	Longest continuous chain of carbon atoms
LFL	Lower flammability limit
LFL (mol%)	Lower flammability limit in mole %
LFL (vol%)	Lower flammability limit in volume %

MAD	Mean absolute deviation
MCC	Microsale combustion calorimetry
MM_i	Molar mass of the repeat unit in polymer
MLR	Multiple linear regression
MNLR	Multivariable nonlinear regression
MNR	Multiple nonlinear regression
MVR	Multi-Variable Regression
MW	Molecular weight
$MW_{repeat\ unit}$	Molecular weight of repeat unit in polymer
n	Optimum number of RBF neurons
n'	Number of carbon atoms under certain conditions given in section 1.2.3.2
n_{Br}	Number of bromine atoms
n'_{Br}	Number of bromine atoms per mole of repeat unit in polymer
n_C	Number of carbon atoms
$n_{C,i}$	Number of carbon atoms in the compound i
n_{Cl}	Number of chlorine atoms
n'_{Cl}	Number of chlorine atoms per mole of repeat unit in polymer
n_F	Number of fluorine atoms
n'_F	Number of fluorine atoms per mole of repeat unit in polymer
n_H	Number of hydrogen atoms
n'_H	Number of hydrogen atoms per mole of repeat unit in polymer
n_I	Number of iodine atoms
n'_N	Number of nitrogen atoms per mole of repeat unit in polymer
n_O	Number of oxygen atoms
n'_O	Number of oxygen atoms per mole of repeat unit in polymer
n_S	Number of sulfur atoms
n_{Si}	Number of silicon atoms
n'_{Si}	Number of silicon atoms per mole of repeat unit in polymer
n_X	Number of halogen atoms
n_i	Number of contribution of functional group i
N_i, M_j, and E_k	Numbers of occurrences of individual group contributions in equation (1.7)
n_{R5}	Number of five member rings in the molecular formula
NBP	Normal boiling point temperature

NRTL	Non-Random Two-Liquid
O_k	Third-order group contributions of the type k in equation (1.7)
OLS	Orthogonal Least Squares
$P^{(+)}$	Increasing function of flash point given in equation (1.15)
$P^{(-)}$	Decreasing function of flash point given in equation (1.15)
PCFC	Pyrolysis combustion flow calorimetry
PCR	Principal component regression
P_i	Partial pressure of compound i in the gas phase above the liquid at $T=$ NBP$_{\text{mix}}$
PLS	Partial least-squares
Q_c^o	Heat of complete combustion of the pyrolysis gases
QSPR	Quantitative structure-property relationship
R	Gas constant
RBF	Radial basis function
rms	Root mean square
SGC	Structural group contribution
SVM	Support vector machine
T_C	Critical temperature
TGA	Thermogravimetric analysis
T_p	Temperature at the peak mass-loss rate in a linear heating program at a constant rate
UFL	Upper flammability limit
UFL (vol%)	Upper flammability limit in volume %
UNIQUAC	UNIversal QUAsiChemical
x_i	Mole fraction of compound i in the liquid phase
X_i	Molar group contributions of char yield of the repeat unit in polymer
y_i	Mole fraction of compound i in the fuel vapor mixture above the liquid at the FP
$\Delta_{\text{vap}}H^o(298.15 \text{ K})$	Standard enthalpy of vaporization at 298.15 K
$\Delta_{\text{vap}}H(\text{at NBP})$	Enthalpy of vaporization at the NBP
φ_i	Amount of contribution of functional group i
v_i	Number of group i of molecule in equations (1.9) and (2.1)
ψ_i	Molar group contributions of heat-release capacity of the repeat unit in polymer
Ω_i	Molar group contributions of total heat release of the repeat unit in polymer
$^1\chi$	First-order molecular connectivity index

σ	Spread parameter
γ_i	Activity coefficient of compound i in the liquid fuel mixture at NBP_{mix}
μ	Weight fraction of the solid residue after pyrolysis or burning
μ^+	Correcting function where it can increase the char yield
μ^-	Correcting function where it can decrease the char yield
$\mu(\%)$	Percent of char yield or the mass fraction of char (g-char/g-sample)
$\left(\Delta H_c^o\right)$	Standard heat of combustion in kJ mol^{-1}
$\Delta H_c^o(\text{gross})$	Gross heat of combustion in kJ mol^{-1}
$\Delta H_c'(\text{gross})$	Gross heat of combustion in kJ kg^{-1}
$\Delta H_c^o(\text{net})$	Net heat of combustion in kJ mol^{-1}
$\Delta H_c'(\text{net})$	Net heat of combustion in kJ kg^{-1}

Appendix A

The reported flash points of combustible organic compounds

No.	Compound	Formula	Exp.	Ref.
1	Methane	CH_4	85	[70]
2	Ethane	C_2H_6	138	[70]
3	Acetylene	C_2H_2	155	[70]
4	Ethylene	C_2H_4	137	[70]
5	Propane	C_3H_8	169	[70]
6	Propene	C_3H_6	165	[70]
7	Propyne	C_3H_4	186	[70]
8	Propadiene	C_3H_4	177	[70]
9	Butane	C_4H_{10}	202	[70]
10	Cyclobutane	C_4H_8	206	[70]
11	1,2-Butadiene	C_4H_6	197	[70]
12	1,3-Butadiene	C_4H_6	197	[70]
13	Butene	C_4H_8	194	[70]
14	cis-2-Butene	C_4H_8	200	[70]
15	Isobutylene	C_4H_8	197	[70]
16	1-Butyne	C_4H_6	207	[70]
17	trans-2-Butene	C_4H_8	200	[70]
18	Cyclobutene	C_4H_6	202	[70]
19	Pentane	C_5H_{12}	224	[70]
20	2,2-Dimethylpropane	C_5H_{14}	208	[70]
21	2-Methylbutane	C_5H_{12}	216	[70]
22	Cyclopentane	C_5H_{10}	231	[70]
23	trans1,2-Dimethylcyclopropane	C_5H_{10}	219	[70]
24	Methylcyclobutane	C_5H_{10}	227	[70]
25	Ethylcyclopropane	C_5H_{10}	231	[70]
26	1-Pentyne	C_5H_8	230	[70]
27	2-Pentyne	C_5H_8	253	[70]
28	3-Methyl-1-butyne	C_5H_8	221	[70]
29	1,2-Pentadiene	C_5H_8	233	[70]
30	2,3-Pentadiene	C_5H_8	235	[70]
31	cis-1,3-Pentadiene	C_5H_8	232	[70]
32	2-Methylbutadiene	C_5H_8	225	[70]
33	3-Methyl-1,2-butadiene	C_5H_8	230	[70]
34	Pentene	C_5H_{10}	229	[70]
35	2-Pentene	C_5H_{10}	253	[70]
36	cis-2-Pentene	C_5H_{10}	227	[70]
37	trans-2-Pentene	C_5H_{10}	225	[70]
38	Isopentene	C_5H_{10}	211	[70]
39	2-Methyl-1-butene	C_5H_{10}	226	[70]
40	1,3-Pentadiene	C_5H_8	244	[70]

(continued)

https://doi.org/10.1515/9783110572223-009

(continued)

No.	Compound	Formula	Exp.	Ref.
41	Cyclopentene	C_5H_8	244	[70]
42	Hexane	C_6H_{14}	250	[70]
43	2,2-Dimethylbutane	C_6H_{14}	225	[70]
44	2,3-Dimethylbutane	C_6H_{14}	244	[70]
45	2-Methylpentane	C_6H_{14}	250	[70]
46	3-Methylpentane	C_6H_{14}	233	[70]
47	Cyclohexane	C_6H_{12}	255	[70]
48	Ethylcyclobutane	C_6H_{12}	250	[70]
49	cis-1-Ethyl-2-methylcyclopropane	C_6H_{12}	244	[70]
50	1-Ethyl-1-methylcyclopropane	C_6H_{12}	240	[70]
51	Benzene	C_6H_6	262	[70]
52	1-Hexyne	C_6H_{10}	252	[70]
53	2-Hexyne	C_6H_{10}	263	[70]
54	3-Hexyne	C_6H_{10}	259	[70]
55	3,3-Dimethyl-1-butyne	C_6H_{10}	239	[70]
56	4-Methyl-1-Pentyne	C_6H_{10}	249	[70]
57	1,4,-Hexadiene	C_6H_{10}	248	[70]
58	2,4-Hexadiene	C_6H_{10}	264	[70]
59	1,5-Hexadiene	C_6H_{10}	246	[70]
60	2,3-Dimethyl-1,3-butadiene	C_6H_{10}	251	[70]
61	3-Methyl-1,4-Pentadiene	C_6H_{10}	239	[70]
62	2-Methyl-2,3-Pentadiene	C_6H_{10}	255	[70]
63	1-Hexene	C_6H_{12}	253	[70]
64	cis-2-Hexene	C_6H_{12}	252	[70]
65	cis-3-Hexene	C_6H_{12}	261	[70]
66	trans-3-Hexene	C_6H_{12}	261	[70]
67	Isohexene	C_6H_{12}	241	[70]
68	2,3-Dimethyl-1-butene	C_6H_{12}	255	[70]
69	2,3-Dimethyl-2-butene	C_6H_{12}	256	[70]
70	3,3-Dimethyl-1-butene	C_6H_{12}	244	[70]
71	2-Methyl-1-penene	C_6H_{12}	241	[70]
72	2-Methyl-2-pentene	C_6H_{12}	246	[70]
73	4-Methyl-2-pentene	C_6H_{12}	241	[70]
74	3-Methyl-1-pentene	C_6H_{12}	244	[70]
75	trans-3-Methyl-2-pentene	C_6H_{12}	266	[70]
76	2-Ethyl-1-butene	C_6H_{12}	243	[70]
77	3-Methyl-1,3-Pentadiene	C_6H_{10}	244	[70]
78	2-Methyl-1,3-pentadiene	C_6H_{10}	255	[70]
79	2-Hexene	C_6H_{12}	253	[70]
80	trans-2-Hexene	C_6H_{12}	246	[70]
81	4,4-Dimethyl-2-pentyne	C_6H_{10}	263	[70]
82	1-Methylcyclopentene	C_6H_{10}	256	[70]
83	Cyclohexene	C_6H_{10}	256	[70]
84	4-Methylcyclopentene	C_6H_{10}	243	[70]

(continued)

No.	Compound	Formula	Exp.	Ref.
85	Heptane	C_7H_{16}	269	[70]
86	3,3-Dimethylpentane	C_7H_{16}	254	[70]
87	2,4-Dimethylpentane	C_7H_{16}	261	[70]
88	3-Ethylpentane	C_7H_{16}	255	[70]
89	2,2-Dimethylpentane	C_7H_{16}	250	[70]
90	2,3-Dimethylpentane	C_7H_{16}	266	[70]
91	2-Methylhexane-	C_7H_{16}	269	[70]
92	3-Methylhexane	C_7H_{16}	258	[70]
93	2,2,3-Trimethylbutane	C_7H_{16}	247	[70]
94	Cycloheptane	C_7H_{14}	279	[70]
95	Methylcyclohexane	C_7H_{14}	269	[70]
96	Ethylcyclopentane	C_7H_{14}	269	[70]
97	trans-1,2-Dimethylcyclopentane	C_7H_{14}	263	[70]
98	Cis-1,2-Dimethylcyclopentane	C_7H_{14}	269	[70]
99	Cis-1,3-Dimethylcyclopentane	C_7H_{14}	259	[70]
100	trans-1,3-Dimethylcyclopentane	C_7H_{14}	260	[70]
101	Toluene	C_7H_8	280	[70]
102	1,6-Heptadiyne	C_7H_8	282	[70]
103	1-Heptyne	C_7H_{12}	263	[70]
104	2-Heptyne	C_7H_{12}	275	[70]
105	3-Heptyne	C_7H_{12}	257	[70]
106	3-Methyl-1-hexyne	C_7H_{12}	268	[70]
107	1,6-Heptadiene	C_7H_{12}	263	[70]
108	1-Heptene	C_7H_{14}	264	[70]
109	cis-2-Heptene	C_7H_{14}	265	[70]
110	trans-2-Heptene	C_7H_{14}	267	[70]
111	trans-3-Heptene	C_7H_{14}	266	[70]
112	2-Methyl-1-hexene	C_7H_{14}	267	[70]
113	4-Methyl-1-hexene	C_7H_{14}	258	[70]
114	2-Ethyl-1-pentene	C_7H_{14}	263	[70]
115	2,4-Dimethyl-2-pentene	C_7H_{14}	264	[70]
116	2,3,3-Trimethyl-1butene	C_7H_{14}	256	[70]
117	cis-5-Methyl-2-Hexene	C_7H_{14}	268	[70]
118	trans-5-Methyl-2-Hexene	C_7H_{14}	268	[70]
119	5-Methyl-1-hexyne	C_7H_{12}	269	[70]
120	2-Methyl-2-hexene	C_7H_{14}	269	[70]
121	3-Methyl-1-hexene	C_7H_{14}	267	[70]
122	3-Ethyl-1-pentene	C_7H_{14}	256	[70]
123	3-Ethyl-2-pentene	C_7H_{14}	267	[70]
124	1-Methylcyclohexene	C_7H_{12}	269	[70]
125	2,4-Dimethyl-1,3-Pentadiene	C_7H_{12}	283	[70]
126	Cycloheptene	C_7H_{12}	267	[70]
127	4-Methylcyclohexene	C_7H_{12}	272	[70]
128	3-Methylcyclohexene	C_7H_{12}	270	[70]

(continued)

(continued)

No.	Compound	Formula	Exp.	Ref.
129	Octane	C_8H_{18}	286	[70]
130	2,2,3,3-Tetramethylbutane	C_8H_{18}	273	[70]
131	2,3,4-Trimethylpentane	C_8H_{16}	273	[70]
132	3,4-Dimethylhexane	C_8H_{18}	277	[70]
133	2,2,4-Trimethylpentane	C_8H_{18}	261	[70]
134	2,2,3-Trimethylpentane	C_8H_{18}	270	[70]
135	3-Methyl-3-ethylpentane	C_8H_{18}	276	[70]
136	3-Ethylhexane	C_8H_{18}	278	[70]
137	3-Methylheptane	C_8H_{18}	279	[70]
138	3,3-Dimethylhexane	C_8H_{18}	272	[70]
139	2,3-Dimethylhexane	C_8H_{18}	283	[70]
140	2,4-Dimethylhexane	C_8H_{18}	283	[70]
141	2,5-Dimethylhexane	C_8H_{18}	271	[70]
142	2,3,3-Trimethylpentane	C_8H_{18}	273	[70]
143	2-Methylheptane	C_8H_{18}	278	[70]
144	2,2-Dimethylhexane	C_8H_{18}	269	[70]
145	4-Methylheptane	C_8H_{18}	278	[70]
146	3-Ethyl-2-methylpentane	C_8H_{18}	276	[70]
147	Cyclooctane	C_8H_{16}	303	[70]
148	Ethylcyclohexane	C_8H_{16}	292	[70]
149	Isopropylcyclopentane	C_8H_{16}	287	[70]
150	n-Propylcyclopentane	C_8H_{16}	289	[70]
151	1,1-Dmethylcyclohexane	C_8H_{16}	276	[70]
152	1-Ethyl-1-methylcyclopentane	C_8H_{16}	279	[70]
153	1,2-Dmethylcyclohexane	C_8H_{16}	288	[70]
154	Cis-1,2-Dmethylcyclohexane	C_8H_{16}	295	[70]
155	trans-1,2-Dmethylcyclohexane	C_8H_{16}	290	[70]
156	1,3-Dmethylcyclohexane	C_8H_{16}	279	[70]
157	Cis-1,3-Dmethylcyclohexane	C_8H_{16}	279	[70]
158	Ethylbenzene	C_8H_{10}	288	[70]
159	P-xylene	C_8H_{10}	300	[70]
160	Trans-1,3-Dmethylcyclohexane	C_8H_{16}	281	[70]
161	Phenylacetylene	C_8H_6	303	[70]
162	Styrene	C_8H_8	304	[70]
163	Phenylacetylene	C_8H_6	303	[70]
164	Styrene	C_8H_8	304	[70]
165	1,7-Octadiyne	C_8H_{10}	296	[70]
166	2,6-Octadiene	C_8H_{10}	307	[70]
167	1-Octyne	C_8H_{14}	289	[70]
168	2-Octyne	C_8H_{14}	301	[70]
169	4-Octyne	C_8H_{14}	291	[70]
170	*trans*-3-Octene	C_8H_{16}	282	[70]
171	*trans*-4-Octene	C_8H_{16}	281	[70]
172	*cis*-4-Octene	C_8H_{16}	294	[70]

(continued)

No.	Compound	Formula	Exp.	Ref.
173	2-Ethyl-1-hexene	C_8H_{16}	279	[70]
174	Cyclooctene	C_8H_{14}	298	[70]
175	1,7-Octadiene	C_8H_{14}	278	[70]
176	1-Octene	C_8H_{16}	281	[70]
177	*trans*-2-Octene	C_8H_{16}	287	[70]
178	6-Methyl-2-heptyne	C_8H_{14}	295	[70]
179	6-Methyl-3-heptyne	C_8H_{14}	289	[70]
180	1,3-Dimethylcyclohexene	C_8H_{14}	285	[70]
181	2,5-Dimethyl-1,5-hexadiene	C_8H_{14}	286	[70]
182	2,5-Dimethyl-2,4-hexadiene	C_8H_{14}	302	[70]
183	2,4,4-Trimethyl-2-pentane	C_8H_{16}	268	[70]
184	2,4,4-Trimethylpentane	C_8H_{16}	256	[70]
185	4-Ethylcyclohexene	C_8H_{14}	286	[70]
186	Nonane	C_9H_{20}	304	[70]
187	2,5-Dimethylheptane	C_9H_{20}	288	[70]
188	3,3-Dimethylheptane	C_9H_{20}	300	[70]
189	3,4-Dimethylheptane	C_9H_{20}	288	[70]
190	2,3-Dimethylheptane	C_9H_{20}	288	[70]
191	2,6-Dimethylheptane	C_9H_{20}	299	[70]
192	2,3-Dimethyl-3-ethylpentane	C_9H_{20}	288	[70]
193	2,2,5-Trimethylhexane	C_9H_{20}	286	[70]
194	4-Methyloctane	C_9H_{20}	295	[70]
195	2-Methyloctane	C_9H_{20}	297	[70]
196	3-Methyloctane	C_9H_{20}	297	[70]
197	2,2,3,4-Tetramethylpentane	C_9H_{20}	284	[70]
198	2,3,3,4-Tetramethylpentane	C_9H_{20}	284	[70]
199	2,2,4,4-Tetramethylpentane	C_9H_{20}	276	[70]
200	3-Ethylheptane	C_9H_{20}	295	[70]
201	4-Ethylheptane	C_9H_{20}	288	[70]
202	2,2-Dimethylheptane	C_9H_{20}	297	[70]
203	2,4-Dimethylheptane	C_9H_{20}	288	[70]
204	2,3,4-Trimethylhexane	C_9H_{20}	288	[70]
205	3,3,4-Trimethylhexane	C_9H_{20}	288	[70]
206	2,3,5-Trimethylhexane	C_9H_{20}	288	[70]
207	2,2,3-Trimethylhexane	C_9H_{20}	288	[70]
208	3,5-Dimethylheptane	C_9H_{20}	288	[70]
209	3,3-Diethylpentane	C_9H_{20}	294	[70]
210	Tetraethylmethane	C_9H_{20}	294	[70]
211	3-Ethyl-2,2-dimethylpentane	C_9H_{20}	286	[70]
212	2,4,4-Trimethylhexane	C_9H_{20}	288	[70]
213	4,4-Dimethylheptane	C_9H_{20}	288	[70]
214	3ethyl-2-methylhexane	C_9H_{20}	288	[70]
215	3-Ethyl-3-methylhexane	C_9H_{20}	288	[70]
216	4-Ethyl-2-methylhexane	C_9H_{20}	288	[70]

(continued)

(continued)

No.	Compound	Formula	Exp.	Ref.
217	3-Ethyl-4-methylhexane	C_9H_{20}	288	[70]
218	2,4-Dimethyl-3-ethylpentane	C_9H_{20}	288	[70]
219	2,2,4-Trimethylhexane	C_9H_{20}	288	[70]
220	2,3,3-Trimethylhexane	C_9H_{20}	288	[70]
221	Cyclononane	C_9H_{18}	316	[70]
222	Isopropylcyclohexane	C_9H_{18}	308	[70]
223	Butylcyclopentane	C_9H_{18}	308	[70]
224	Propylcyclohexane	C_9H_{18}	304	[70]
225	1-Nonyne	C_9H_{16}	306	[70]
226	1,8-Nonadiene	C_9H_{16}	299	[70]
227	Propylbenzene	C_9H_{12}	303	[70]
228	Cumene	C_9H_{12}	304	[70]
229	m-Ethyltoluene	C_9H_{12}	311	[70]
230	1,2,3-Trimethylbenzene	C_9H_{12}	324	[70]
231	1,2,4-Trimethylbenzene	C_9H_{12}	321	[70]
232	1,3,5-Trimethylbenzene	C_9H_{12}	317	[70]
233	o-Ethyltoluene	C_9H_{12}	312	[70]
234	p-Ethyltoluene	C_9H_{12}	309	[70]
235	1,8-Nonadiyne	C_9H_{12}	314	[70]
236	2-Vinyltoluene	C_9H_{10}	320	[70]
237	3-vinyltoluene	C_9H_{10}	324	[70]
238	Allylbenzene	C_9H_{10}	310	[70]
239	beta-Methylstyrene	C_9H_{10}	333	[70]
240	cis-1-Propenylbenzene	C_9H_{10}	325	[70]
241	Isopropenylbenzene	C_9H_{10}	313	[70]
242	trans-1-phenyl-1-propene	C_9H_{10}	331	[70]
243	4-Vinyltoluene	C_9H_{10}	318	[70]
244	Decane	$C_{10}H_{22}$	319	[70]
245	3-Ethyloctane	$C_{10}H_{22}$	314	[70]
246	4-Ethyloctane	$C_{10}H_{22}$	314	[70]
247	2,2,4-Trimethylheptane	$C_{10}H_{22}$	304	[70]
248	3,4,4-Trimethylheptane	$C_{10}H_{22}$	304	[70]
249	3,3,4-Trimethylheptane	$C_{10}H_{22}$	304	[70]
250	2,4,4-Trimethylheptane	$C_{10}H_{22}$	304	[70]
251	2,3,6-Trimethylheptane	$C_{10}H_{22}$	304	[70]
252	2,3,5-Trimethylheptane	$C_{10}H_{22}$	304	[70]
253	2,3,3-Trimethylheptane	$C_{10}H_{22}$	304	[70]
254	2,2,5-Trimethylheptane	$C_{10}H_{22}$	304	[70]
255	2,2,6-Trimethylheptane	$C_{10}H_{22}$	304	[70]
256	2,4,5-Trimethylheptane	$C_{10}H_{22}$	304	[70]
257	2,5-Dimethyloctane	$C_{10}H_{22}$	314	[70]
258	2,4-Dimethyloctane	$C_{10}H_{22}$	314	[70]
259	4,4-Dimethyloctane	$C_{10}H_{22}$	314	[70]
260	4,5-Dimethyloctane	$C_{10}H_{22}$	314	[70]

(continued)

No.	Compound	Formula	Exp.	Ref.
261	3,4-Dimethyloctane	$C_{10}H_{22}$	314	[70]
262	3,6-Dimethyloctane	$C_{10}H_{22}$	314	[70]
263	2,3-Dimethyloctane	$C_{10}H_{22}$	314	[70]
264	3,3-Diethylhexane	$C_{10}H_{22}$	311	[70]
265	3,3-Dimethyloctane	$C_{10}H_{22}$	314	[70]
266	3,5-Dimethyloctane	$C_{10}H_{22}$	314	[70]
267	2,6-Dimethyloctane	$C_{10}H_{22}$	314	[70]
268	3-Ethyl-2,2-dimethylhexane	$C_{10}H_{22}$	311	[70]
269	3,3,4,4-Tetramethylhexane	$C_{10}H_{22}$	304	[70]
270	2,2,5,5-Tetramethylhexane	$C_{10}H_{22}$	304	[70]
271	2,2,3,5-Tetramethylhexane	$C_{10}H_{22}$	304	[70]
272	2,3,3,4-Tetramethylhexane	$C_{10}H_{22}$	304	[70]
273	2,3,3,5-Tetramethylhexane	$C_{10}H_{22}$	304	[70]
274	2,2,4,5-Tetramethylhexane	$C_{10}H_{22}$	304	[70]
275	2,3,4,5-Tetramethylhexane	$C_{10}H_{22}$	304	[70]
276	2,2,4,4-Tetramethylhexane	$C_{10}H_{22}$	304	[70]
277	3,3,5-Trimethylheptane	$C_{10}H_{22}$	304	[70]
278	2,3,5-Trimethylheptane	$C_{10}H_{22}$	304	[70]
279	3-Ethyl-3-Methylheptan	$C_{10}H_{22}$	314	[70]
280	5-Ethyl-2-Methylheptan	$C_{10}H_{22}$	314	[70]
281	3-Ethyl-4-Methylheptan	$C_{10}H_{22}$	304	[70]
282	4-Ethyl-3-Methylheptan	$C_{10}H_{22}$	314	[70]
283	4-Methylnonane	$C_{10}H_{22}$	311	[70]
284	3-Methylnonane	$C_{10}H_{22}$	314	[70]
285	2-Methylnonane	$C_{10}H_{22}$	314	[70]
286	2,4,6-Trimethylheptane	$C_{10}H_{22}$	304	[70]
287	3-Ethyl-2,3,4-Trimethylpentane	$C_{10}H_{22}$	304	[70]
288	2,3,4,4-Tetramethylhexane	$C_{10}H_{22}$	304	[70]
289	3,4,5-Trimethylheptane	$C_{10}H_{22}$	304	[70]
290	3-Ethyl-5-methylheptane	$C_{10}H_{22}$	304	[70]
291	5-Methylnonane	$C_{10}H_{22}$	312	[70]
292	2,2,3,4-Tetramethylhexane	$C_{10}H_{22}$	304	[70]
293	4-Propylheptane	$C_{10}H_{22}$	314	[70]
294	Cyclodecane	$C_{10}H_{20}$	338	[70]
295	n-Butylcyclohexane	$C_{10}H_{20}$	314	[70]
296	Pentylcyclopentane	$C_{10}H_{20}$	324	[70]
297	Butylbenzene	$C_{10}H_{14}$	331	[70]
298	1,2,4,5-Tetramethylbenzene	$C_{10}H_{14}$	346	[70]
299	2-Ethyl-p-xylene	$C_{10}H_{14}$	329	[70]
300	3-Ethyl-o-xylene	$C_{10}H_{14}$	338	[70]
301	4-Ethyl-m-xylene	$C_{10}H_{14}$	330	[70]
302	tert-Butylbenzene	$C_{10}H_{14}$	307	[70]
303	P-Cymene	$C_{10}H_{14}$	320	[70]
304	o-DiEthylbenzene	$C_{10}H_{14}$	322	[70]

(continued)

(continued)

No.	Compound	Formula	Exp.	Ref.
305	m-DiEthylbenzene	$C_{10}H_{14}$	324	[70]
306	p-DiEthylbenzene	$C_{10}H_{14}$	328	[70]
307	4-Ethyl-1,2-dimethylbenzene	$C_{10}H_{14}$	331	[70]
308	m-Divinylbenzene	$C_{10}H_{10}$	338	[70]
309	p-Divinylbenzene	$C_{10}H_{10}$	337	[70]
310	1-Butynylbenzene	$C_{10}H_{10}$	341	[70]
311	Naphthalene	$C_{10}H_8$	360	[70]
312	3-Ethylstyrene	$C_{10}H_{12}$	333	[70]
313	4-Ethylstyrene	$C_{10}H_{12}$	335	[70]
314	2,4-Dimethylstyrene	$C_{10}H_{12}$	333	[70]
315	1-Decyne	$C_{10}H_{18}$	323	[70]
316	sec-Butylbenzene	$C_{10}H_{14}$	318	[70]
317	Isobutylbenzene	$C_{10}H_{14}$	323	[70]
318	1,2,3,4-Tetramethylbenzene	$C_{10}H_{14}$	341	[70]
319	1,2,3,5-Tetramethylbenzene	$C_{10}H_{14}$	336	[70]
320	m-Cymene	$C_{10}H_{14}$	323	[70]
321	p-Methylisopropenylbenzene	$C_{10}H_{12}$	332	[70]
322	cis-(1-Methyl-1-propenyl)benzene	$C_{10}H_{12}$	328	[70]
323	1,9-Decadiene	$C_{10}H_{18}$	314	[70]
324	1-Decene	$C_{10}H_{20}$	322	[70]
325	trans-5-Decene	$C_{10}H_{20}$	319	[70]
326	3-Menthene	$C_{10}H_{18}$	316	[70]
327	trans-1,2-Di-tert-butylethylene	$C_{10}H_{20}$	306	[70]
328	3,7-Dimethyl-1-octene	$C_{10}H_{20}$	317	[70]
329	Undecane	$C_{11}H_{24}$	333	[70]
330	Hexylcyclopentane	$C_{11}H_{22}$	339	[70]
331	Pentylcyclohexane	$C_{11}H_{22}$	339	[70]
332	1-methylnaphtalene	$C_{11}H_{10}$	355	[70]
333	n-Pentylbenzene	$C_{11}H_{16}$	339	[70]
334	IsoPentylbenzene	$C_{11}H_{16}$	335	[70]
335	Pentamethylbenzene	$C_{11}H_{16}$	364	[70]
336	p-tert-Butyltoluene	$C_{11}H_{16}$	321	[70]
337	2-Phenyl-2methylbutane	$C_{11}H_{16}$	338	[70]
338	1-Undecyne	$C_{11}H_{20}$	338	[70]
339	4-Undecyne	$C_{11}H_{20}$	341	[70]
340	1-Undecene	$C_{11}H_{22}$	336	[70]
341	Neopentylbenzene	$C_{11}H_{16}$	323	[70]
342	2-Methylnaphtalene	$C_{11}H_{10}$	371	[70]
343	3,4-Dimethylcumene	$C_{11}H_{16}$	341	[70]
344	1-Pentynylbenzene	$C_{11}H_{12}$	355	[70]
345	Dodecane	$C_{12}H_{26}$	344	[70]
346	Cyclododecane	$C_{12}H_{24}$	371	[70]
347	1-Cyclohexylhexane	$C_{12}H_{24}$	353	[70]
348	Heptylcyclopentane	$C_{12}H_{24}$	352	[70]

(continued)

No.	Compound	Formula	Exp.	Ref.
349	1-Ethylnaphtalene	$C_{12}H_{12}$	380	[70]
350	2-Ethylnaphtalene	$C_{12}H_{12}$	377	[70]
351	1,3-Dimethylnaphtalene	$C_{12}H_{12}$	382	[70]
352	1,2-Dimethylnaphtalene	$C_{12}H_{12}$	374	[70]
353	Hexylbenzene	$C_{12}H_{18}$	356	[70]
354	Hexamethylbenzene	$C_{12}H_{18}$	377	[70]
355	3,5-Dimethyl-tert-butylbenzene	$C_{12}H_{18}$	357	[70]
356	1,2,4-Trimethylbenzene	$C_{12}H_{18}$	349	[70]
357	1,3,5-Trimethylbenzene	$C_{12}H_{18}$	354	[70]
358	1,4-Diisopropylbenzene	$C_{12}H_{18}$	354	[70]
359	m-Diisopropylbenzene	$C_{12}H_{18}$	350	[70]
360	1-Dodecyne	$C_{12}H_{22}$	352	[70]
361	Dodecane	$C_{12}H_{24}$	351	[70]
362	2-Methyl-1-undecene	$C_{12}H_{24}$	345	[70]
363	1,4-Dimethylnaphtalene	$C_{12}H_{12}$	383	[70]
364	2,3-Dimethylnaphtalene	$C_{12}H_{12}$	387	[70]
365	Diisopropylbenzene	$C_{12}H_{18}$	344	[70]
366	1-Vinylnaphthalene	$C_{12}H_{10}$	389	[70]
367	1,5-Dimethylnaphtalene	$C_{12}H_{12}$	384	[70]
368	1,6-Dimethylnaphtalene	$C_{12}H_{12}$	383	[70]
369	2,7-Dimethylnaphtalene	$C_{12}H_{12}$	382	[70]
370	2,6-Dimethylnaphtalene	$C_{12}H_{12}$	382	[70]
371	Tridecane	$C_{13}H_{28}$	352	[70]
372	1-Cyclohexylheptane	$C_{13}H_{26}$	366	[70]
373	Octylcyclopentane	$C_{13}H_{26}$	365	[70]
374	n-Heptylbenzene	$C_{13}H_{20}$	368	[70]
375	1-Tridecyne	$C_{13}H_{24}$	366	[70]
376	1-Tridecene	$C_{13}H_{26}$	352	[70]
377	Tetradecane	$C_{14}H_{30}$	372	[70]
378	Cyclohexyloctane	$C_{14}H_{28}$	378	[70]
379	1,2,3,4-Tetraethylbenzene	$C_{14}H_{22}$	367	[70]
380	2-Phenyloctane	$C_{14}H_{22}$	373	[70]
381	n-Octylbenzene	$C_{14}H_{22}$	380	[70]
382	1-Tetradecene	$C_{14}H_{28}$	383	[70]
383	1,4-Di-tert-butylbenzene	$C_{14}H_{22}$	370	[70]
384	Tetradecyne	$C_{14}H_{26}$	378	[70]
385	Cyclohexylnonane	$C_{15}H_{30}$	390	[70]
386	n-Nonylbenzene	$C_{15}H_{24}$	390	[70]
387	1,3,5-Triisopropylbenzene	$C_{15}H_{24}$	359	[70]
388	1-Methylanthracene	$C_{15}H_{12}$	430	[70]
389	2-Methylanthracene	$C_{15}H_{12}$	431	[70]
390	9-Methylanthracene	$C_{15}H_{12}$	431	[70]
391	1-methylphenathrene	$C_{15}H_{12}$	431	[70]
392	1-Pentadecene	$C_{15}H_{30}$	386	[70]

(continued)

(continued)

No.	Compound	Formula	Exp.	Ref.
393	1-Pentadecyne	$C_{15}H_{28}$	390	[70]
394	Hexadecane	$C_{16}H_{34}$	408	[70]
395	Cyclopentylundecane	$C_{16}H_{32}$	399	[70]
396	Cyclohexyldecane	$C_{16}H_{32}$	404	[70]
397	Decylbenzene	$C_{16}H_{26}$	380	[70]
398	Pentaethylbenzene	$C_{16}H_{26}$	386	[70]
399	1-Hexadecene	$C_{16}H_{32}$	402	[70]
400	9,10-Dimethylanthracene	$C_{16}H_{14}$	442	[70]
401	1-Hexadecyne	$C_{16}H_{30}$	401	[70]
402	9-Vinylanthracene	$C_{16}H_{12}$	445	[70]
403	1-Phenylnaphthalene	$C_{16}H_{12}$	435	[70]
404	Heptadecane	$C_{17}H_{36}$	421	[70]
405	Undecylcyclohexane	$C_{17}H_{34}$	411	[70]
406	Cyclopentyldodecane	$C_{17}H_{34}$	409	[70]
407	n-Undecylbenzene	$C_{17}H_{28}$	409	[70]
408	1-Heptadecene	$C_{17}H_{34}$	408	[70]
409	Octadecane	$C_{18}H_{38}$	438	[70]
410	Cyclohexyldodecane	$C_{18}H_{36}$	420	[70]
411	Cyclopentyltridecane	$C_{18}H_{36}$	418	[70]
412	Retene	$C_{18}H_{18}$	451	[70]
413	1-Octadecene	$C_{18}H_{36}$	421	[70]
414	1-Octadecyne	$C_{18}H_{34}$	422	[70]
415	Cyclohexyltridecane	$C_{19}H_{38}$	429	[70]
416	Tridecylbenzene	$C_{19}H_{32}$	385	[70]
417	Cyclopentyltetradecane	$C_{19}H_{38}$	427	[70]
418	1-Nonadeyne	$C_{19}H_{36}$	431	[70]
419	1-Nonadecene	$C_{19}H_{38}$	430	[70]
420	1-Eicosene	$C_{20}H_{40}$	439	[70]
421	Methylamine	CH_5N	215.0	[43]
422	Dicyandiamide	$C_2H_4N_4$	474.0	[43]
423	Monoethanolamine	C_2H_7N	358.2	[43]
424	Dimethylamine	C_2H_7N	223.2	[43]
425	Ethylenediamine	$C_2H_8N_2$	306.2	[43]
426	Ethyleneimine	C_2H_5N	262.0	[43]
427	Ethylamine	C_2H_7N	228.0	[43]
428	Allylamine	C_3H_7N	244.3	[43]
429	Cyclopropylamine	C_3H_7N	245.0	[43]
430	Isopropylamine	C_3H_9N	236.2	[43]
431	Melamine	$C_3H_6N_6$	570.0	[43]
432	Methylethanolamine	C_3H_9N	345.2	[43]
433	1,3-Propanediamine	$C_3H_{10}N_2$	321.2	[43]
434	1,2-Propanediamine	$C_3H_{10}N_2$	306.2	[43]
435	n-Propylamine	C_3H_9N	261.0	[43]
436	Oxazole	C_3H_3NO	292.0	[43]

(continued)

No.	Compound	Formula	Exp.	Ref.
437	Propyleneimine	C_3H_7N	263.2	[43]
438	Pyrazole	$C_3H_4N_2$	355.0	[43]
439	Trimethylamine	C_3H_9N	266.0	[43]
440	n-Aminoethyl ethanolamine	$C_4H_{12}N_2O$	375.0	[43]
441	Bis-(2-aminoethyl)ether	$C_4H_{12}N_2O$	360.0	[43]
442	sec-Butylamine	$C_4H_{11}N$	244.3	[43]
443	n-Butylamine	$C_4H_{11}N$	261.0	[43]
444	tert-Butylamine	$C_4H_{11}N$	264.2	[43]
445	Diethanolamine	$C_4H_{11}NO_2$	425.0	[43]
446	Diethylamine	$C_4H_{11}N$	234.2	[43]
447	Diethylenetriamine	$C_4H_{13}N_3$	372.0	[43]
448	n,n-Diethylhydroxylamine	$C_4H_{11}N$	318.2	[43]
449	Dimethylethanolamine	$C_4H_{11}N$	314.2	[43]
450	Piperazine	$C_4H_{10}N_2$	354.3	[43]
451	Pyrrole	C_4H_5N	312.0	[43]
452	Pyrrolidine	C_4H_9N	276.0	[43]
453	2-Pyrrolidone	C_4H_7NO	402.0	[43]
454	Morpholine	C_4H_9NO	308.2	[43]
455	Isobutylamine	$C_4H_{11}N$	255.4	[43]
456	3-Methoxyisopropylamine	$C_4H_{11}N$	281.2	[43]
457	Cyclopentylamine	$C_5H_{11}N$	290.2	[43]
458	n,n-Diethylmethylamine	$C_5H_{13}N$	250.0	[43]
459	3-(n,n-Dimethylamino)propylamine	$C_5H_{14}N_2$	304.0	[43]
460	1,5-Pentanediamine	$C_5H_{14}N_2$	335.2	[43]
461	n-Pentylamine	$C_5H_{13}N$	272.2	[43]
462	Piperidine	$C_5H_{11}N$	276.2	[43]
463	4-Formylmorpholine	$C_5H_9NO_2$	386.2	[43]
464	n-Methylpyrrole	C_5H_7N	288.2	[43]
465	n-Methylpyrrolidine	$C_5H_{11}N$	259.0	[43]
466	n-Methyl-2-pyrrolidone	C_5H_9N	359.2	[43]
467	Methyl diethanolamine	$C_5H_{13}NO_2$	399.8	[43]
468	2-Methyl-2-aminobutane	$C_5H_{13}N$	257.0	[43]
469	4-(2-Aminoethyl)morpholine	$C_6H_{14}N_2O$	359.2	[43]
470	n-Aminoethyl piperazine	$C_6H_{15}N_3O_{10}$	366.2	[43]
471	Aniline	C_6H_7N	343.2	[43]
472	m-Chloroaniline	C_6H_6N	378.0	[43]
473	p-Chloroaniline	C_6H_6N	386.0	[43]
474	o-Chloroaniline	C_6H_6N	363.7	[43]
475	Cyclohexylamine	$C_6H_{13}N$	304.2	[43]
476	3,4-Dichloroaniline	C_6H_5N	439.0	[43]
477	Diethylethanolamine	$C_6H_{15}N$	321.2	[43]
478	Diisopropylamine	$C_6H_{15}N$	266.2	[43]
479	n,n-Dimethyl-n-butylamine	$C_6H_{15}N$	270.2	[43]
480	n-Ethyl-2-methylallylamine	$C_6H_{13}N$	280.2	[43]

(continued)

(continued)

No.	Compound	Formula	Exp.	Ref.
481	Hexamethylenediamine	$C_6H_{16}N_2$	366.0	[43]
482	Hexamethyleneimine	$C_6H_{13}N$	310.0	[43]
483	Hexamethylenetetramine	$C_6H_{12}N_4$	523.2	[43]
484	n-Hexylamine	$C_6H_{15}N$	281.2	[43]
485	2,2π,2γPrime;-nitrilotrisacetonitrile	$C_6H_6N_4$	487.0	[43]
486	m-Nitroaniline	$C_6H_6N_2O_2$	472.2	[43]
487	o-Nitroaniline	$C_6H_6N_2O_2$	441.2	[43]
488	p-Nitroaniline	$C_6H_6N_2O_2$	472.2	[43]
489	m-Phenylenediamine	$C_6H_8N_2$	411.2	[43]
490	p-Phenylenediamine	$C_6H_8N_2$	428.7	[43]
491	o-Phenylenediamine	$C_6H_8N_2$	429.2	[43]
492	Phenylhydrazine	$C_6H_8N_2$	362.2	[43]
493	Di-n-propylamine	$C_6H_{15}N$	290.2	[43]
494	Tetramethylethylenediamine	$C_6H_{16}N_2$	283.2	[43]
495	Triethanolamine	$C_6H_{15}NO_3$	453.0	[43]
496	Triethylamine	$C_6H_{15}N$	258.2	[43]
497	Triethylenediamine	$C_6H_{12}N_2$	335.0	[43]
498	Triethylenetetramine	$C_6H_{18}N_4$	408.2	[43]
499	n-Ethylmorpholine	C_6H_3NO	303.2	[43]
500	n-(2-Hydroxyethyl)piperazine	$C_6H_{14}N_2O$	397.2	[43]
501	2-Methylpyridine	C_6H_7N	299.2	[43]
502	4-Methylpyridine	C_6H_7N	329.2	[43]
503	3-Methylpyridine	C_6H_7N	309.2	[43]
504	Nicotinonitrile	$C_6H_4N_2$	357.2	[43]
505	n-Heptylamine	$C_7H_{17}N$	308.2	[43]
506	Benzylamine	C_7H_9N	333.2	[43]
507	2,6-Diaminotoluene	$C_7H_{10}N$	419.0	[43]
508	n-Methylaniline	C_7H_9N	351.0	[43]
509	n-Methylcyclohexylamine	$C_7H_{15}N$	302.2	[43]
510	m-Toluidine	C_7H_9N	359.2	[43]
511	o-Toluidine	C_7H_9N	358.2	[43]
512	p-Toluidine	C_7H_9N	360.0	[43]
513	2,6-Dimethylpyridine	C_7H_9N	306.0	[43]
514	4-[2-(2-Aminoethoxy)ethyl]morpholine	$C_8H_{18}N_2O_2$	388.0	[43]
515	Di-n-butylamine	$C_8H_{19}N$	312.2	[43]
516	Diisobutylamine	$C_8H_{19}N$	294.2	[43]
517	n,n-Dimethylaniline	$C_8H_{11}N$	335.2	[43]
518	2,4-Dimethylaniline	$C_8H_{11}N$	371.2	[43]
519	o-Ethylaniline	$C_8H_{11}N$	364.2	[43]
520	n-Ethylaniline	$C_8H_{11}N$	358.2	[43]
521	n-Octylamine	$C_8H_{19}N$	333.0	[43]
522	p-Phenetidine	$C_8H_{11}N$	389.0	[43]
523	Tetraethylenepentamine	$C_8H_{23}N_5$	436.0	[43]
524	2-(n-Morpholino)-2π-hydroxydiethylether	$C_8H_{17}NO_3$	403.0	[43]

(continued)

No.	Compound	Formula	Exp.	Ref.
525	n-Nonylamine	$C_9H_{21}N$	335.2	[43]
526	Triallylamine	$C_9H_{15}N$	304.0	[43]
527	Tripropylamine	$C_9H_{21}N$	302.0	[43]
528	p-Dimethylaminobenzaldehyde	$C_9H_{11}NO$	420.2	[43]
529	n,nπ-Di-tert-butylethylenediamine	$C_{10}H_{24}N_2$	334.0	[43]
530	n-Decylamine	$C_{10}H_{23}N$	358.2	[43]
531	Diamylamine	$C_{10}H_{23}N$	325.2	[43]
532	2,6-Diethylaniline	$C_{10}H_{15}N$	396.2	[43]
533	n,n-Diethylaniline	$C_{10}H_{15}N$	358.2	[43]
534	Pentaethylene hexamine	$C_{10}H_{28}N_6$	509.0	[43]
535	Undecylamine	$C_{11}H_{25}N$	365.2	[43]
536	p-Aminodiphenylamine	$C_{12}H_{12}N_2$	466.5	[43]
537	Benzidine	$C_{12}H_{12}N_2$	521.0	[43]
538	Tri-n-butylamine	$C_{12}H_{27}N$	359.3	[43]
539	Dicyclohexylamine	$C_{12}H_{23}N$	377.0	[43]
540	4,4π-Dinitrodiphenylamine	$C_{12}H_9N_3O_4$	544.0	[43]
541	Diphenylamine	$C_{12}H_{11}N$	426.0	[43]
542	n-Dodecylamine	$C_{12}H_{27}N$	382.0	[43]
543	Hydrazobenzene	$C_{12}H_{12}N_2$	503.0	[43]
544	n-Tetradecylamine	$C_{14}H_{31}N$	406.0	[43]
545	Triamylamine	$C_{15}H_{33}N$	376.0	[43]
546	Di-2-ethylhexylamine	$C_{16}H_{35}N$	405.0	[43]
547	di-n-Octylamine	$C_{16}H_{35}N$	383.2	[43]
548	n,nπ-Diphenyl-pphenylenediamine	$C_{18}H_{16}N$	505.2	[43]
549	4,4π-Dinitrotriphenylamine	$C_{18}H_{13}N_3O_4$	557.0	[43]
550	Dehydroabietylamine	$C_{20}H_{31}N$	464.0	[43]
551	tri-n-Octylamine	$C_{24}H_{51}N$	457.0	[43]
552	Hydrazine	H_4N_2	311.0	[43]
553	Methanol	CH_4O	288.8	[71]
554	2,2,2-Trifluoroethanol	$C_2H_3F_3O$	314.2	[71]
555	2,2-Dichloroethanol	$C_2H_4Cl_2O$	351.5	[71]
556	2-Fluoroethanol	C_2H_5FO	307.2	[71]
557	2-Chloroethanol	C_2H_5ClO	333.2	[255]
558	2-Bromoethanol	C_2H_5BrO	313.7	[71]
559	2-Iodoethanol	C_2H_5IO	338.7	[71]
560	Ethanol	C_2H_6O	292.1	[71]
561	2-Mercaptoethanol	C_2H_6OS	350.2	[71]
562	1,2-Ethanediol (or Ethylene glycol)	$C_2H_6O_2$	391.5	[71]
563	2-Aminoethanol (or Ethanolamine)	C_2H_7NO	366.5	[71]
564	2-propyn-1-ol (or propargyl alcohol)	C_3H_4O	309.3	[256]
565	2,3-Dibromo-2-propen-1-ol	$C_3H_4Br_2O$	377.2	[71]
566	1,1,1-Trichloro-2-propanol	$C_3H_5Cl_3O$	355.4	[71]
567	2-propen-1-ol (or allyl alcohol)	C_3H_6O	305.4	[71]
568	1,3-Difluoro-2-propanol	$C_3H_6F_2O$	315.4	[71]

(continued)

(continued)

No.	Compound	Formula	Exp.	Ref.
569	1,3-Dichloro-2-propanol	$C_3H_6Cl_2O$	358.2	[165]
570	2,3-Dichloro-1-propanol	$C_3H_6Cl_2O$	366.5	[71]
571	1,3-Dibromo-2-propanol	$C_3H_6Br_2O$	319.8	[71]
572	1-Hydroxy-2-propanone (or Acetol)	$C_3H_6O_2$	329.3	[165]
573	2,3-Epoxy-1-propanol (or Glycidol)	$C_3H_6O_2$	354.3	[71]
574	1-Chloro-2-propanol	C_3H_7ClO	324.8	[165]
575	1-Chloro-3-propanol	C_3H_7ClO	346.2	[165]
576	3-Bromo-1-propanol	C_3H_7BrO	367.0	[71]
577	1-Bromo-2-propanol	C_3H_7BrO	316.2	[71]
578	3-Chloro-1,2-propanediol	$C_3H_7ClO_2$	410.9	[71]
579	3-Bromo-1,2-propanediol	$C_3H_7BrO_2$	383.2	[71]
580	2-Nitro-1-propanol	$C_3H_7NO_3$	373.2	[71]
581	1-Propanol	C_3H_8O	305.4	[71]
582	2-Propanol (or Isopropanol)	C_3H_8O	294.3	[71]
583	2-(Methylthio)ethanol	C_3H_8OS	343.2	[71]
584	2-Methoxyethanol	$C_3H_8O_2$	325.2	[71]
585	1,2-Propanediol (or 1,2-Propylene glycol)	$C_3H_8O_2$	380.4	[71]
586	1,3-Propanediol (or 1,3-Propylene glycol)	$C_3H_8O_2$	352.6	[71]
587	1,2,3-Propanetriol (or Glycerol)	$C_3H_8O_3$	433.0	[256]
588	1-Amino-2-propanol	C_3H_9NO	350.4	[71]
589	2-Amino-1-propanol	C_3H_9NO	357.0	[71]
590	3-Amino-1-propanol	C_3H_9NO	353.2	[255]
591	2-(Methylamino) ethanol	C_3H_9NO	363.2	[71]
592	1,3-Diamino-2-propanol	$C_3H_{10}N_2O$	405.4	[71]
593	2,2,3,3,4,4,4-Heptafluoro-1-butanol	$C_4H_3F_7O$	293.2	[71]
594	2-Butyn-1-ol	C_4H_6O	324.8	[71]
595	3-Butyn-1-ol	C_4H_6O	309.3	[71]
596	3-Butyn-2-ol	C_4H_6O	307.6	[71]
597	2-Butyne-1,4-diol	$C_4H_6O_2$	425.4	[71]
598	2-Hydroxy-2-methylpropanenitrile	C_4H_7NO	336.2	[165]
599	1-Ethoxy-2,2,2-trifluoro-1-ethanol	$C_4H_7F_3O_2$	312.6	[71]
600	3-Buten-1-ol	C_4H_8O	310.9	[71]
601	2-Buten-1-ol	C_4H_8O	318.1	[71]
602	3-Buten-2-ol	C_4H_8O	289.8	[71]
603	Cyclopropylmethanol (or Cyclopropyl carbinol)	C_4H_8O	313.2	[71]
604	Cyclobutanol	C_4H_8O	294.3	[71]
605	3-Hydroxy-2-butanone (or Acetyl methyl carbinol)	$C_4H_8O_2$	323.2	[71]
606	3-Hydroxybutanal	$C_4H_8O_2$	339.0	[165]
607	(cis) 2-Butene-1,4-diol	$C_4H_8O_2$	401.5	[71]
608	(trans) 2-Butene-1,4-diol	$C_4H_8O_2$	382.2	[71]
609	1-Aziridineethanol	C_4H_9NO	340.4	[71]
610	1-Chloro-2-methyl-2-propanol	C_4H_9ClO	311.2	[71]
611	2-(2-Chloroethoxy) ethanol	$C_4H_9ClO_2$	363.7	[71]

(continued)

No.	Compound	Formula	Exp.	Ref.
612	3-Nitro-2-butanol	$C_4H_9NO_3$	364.3	[71]
613	1-Butanol	$C_4H_{10}O$	319.3	[71]
614	2-Butanol (or sec-Butanol)	$C_4H_{10}O$	304.2	[71]
615	2-Methyl-1-propanol (or Isobutanol)	$C_4H_{10}O$	312.6	[71]
616	2-Methyl-2-propanol (or tert-Butanol)	$C_4H_{10}O$	288.7	[71]
617	3-Mercapto-2-butanol	$C_4H_{10}OS$	334.8	[71]
618	1,2-Butanediol	$C_4H_{10}O_2$	366.5	[71]
619	1,3-Butanediol	$C_4H_{10}O_2$	394.3	[71]
620	1,4-Butanediol	$C_4H_{10}O_2$	394.3	[71]
621	2,3-Butanediol	$C_4H_{10}O_2$	358.2	[71]
622	2-Methoxy-1-propanol	$C_4H_{10}O_2$	309.0	[165]
623	2-Ethoxyethanol	$C_4H_{10}O_2$	322.2	[71]
624	1-Methoxy-2-propanol	$C_4H_{10}O_2$	311.2	[71]
625	2,2′-Thiodiethanol	$C_4H_{10}O_2S$	433.2	[71]
626	2,2′-Oxybis(ethanol) (or Diethylene glycol)	$C_4H_{10}O_3$	397.2	[255]
627	1,2,3-Butanetriol	$C_4H_{10}O_3$	419.0	[71]
628	1,2,4-Butanetriol	$C_4H_{10}O_3$	439.8	[71]
629	2-(Hydroxymethyl)-1,3-propanediol (or Trimethylolmethane)	$C_4H_{10}O_3$	426.9	[71]
630	2-Amino-1-butanol	$C_4H_{11}NO$	368.2	[71]
631	4-Amino-1-butanol	$C_4H_{11}NO$	380.9	[71]
632	2-Amino-2-methyl-1-propanol	$C_4H_{11}NO$	356.5	[71]
633	N,N-Dimethylethanolamine	$C_4H_{11}NO$	314.2	[71]
634	2-(Ethylamino)ethanol	$C_4H_{11}NO$	345.4	[71]
635	2-(2-Aminoethoxy)ethanol	$C_4H_{11}NO_2$	397.0	[165]
636	Diethanolamine	$C_4H_{11}NO_2$	445.4	[71]
637	2-(2-Aminoethylamino)ethanol (or n-Aminoethyl ethanolamine)	$C_4H_{12}N_2O$	408.2	[71]
638	2-Furanylmethanol (or Furfuryl alcohol)	$C_5H_6O_2$	350.2	[71]
639	3-Furanylmethanol	$C_5H_6O_2$	311.5	[71]
640	2-Methyl-3-butyn-2-ol	C_5H_8O	303.9	[71]
641	3-Pentyn-1-ol	C_5H_8O	327.6	[71]
642	4-Pentyn-2-ol	C_5H_8O	310.4	[71]
643	1-Pentyn-3-ol (or Ethyl ethynyl carbinol)	C_5H_8O	302.6	[71]
644	3,3,4,4-Tetrafluoro-2-methyl-2-butanol	$C_5H_8F_4O$	347.0	[71]
645	1-Methylcyclopropanemethanol	$C_5H_{10}O$	307.0	[71]
646	2-Methylcyclopropanemethanol	$C_5H_{10}O$	312.6	[71]
647	Cyclopropylethanol	$C_5H_{10}O$	310.2	[71]
648	Cyclobutanemethanol	$C_5H_{10}O$	313.2	[71]
649	Cyclopentanol	$C_5H_{10}O$	324.3	[71]
650	2-Methyl-3-buten-2-ol	$C_5H_{10}O$	286.5	[71]
651	3-Methyl-2-buten-1-ol	$C_5H_{10}O$	316.5	[71]
652	3-Methyl-3-buten-1-ol	$C_5H_{10}O$	315.2	[71]
653	1-Penten-3-ol	$C_5H_{10}O$	301.2	[71]

(continued)

(continued)

No.	Compound	Formula	Exp.	Ref.
654	2-Penten-1-ol	$C_5H_{10}O$	323.7	[71]
655	(cis) 2-Penten-1-ol	$C_5H_{10}O$	321.5	[71]
656	3-Penten-2-ol	$C_5H_{10}O$	309.2	[71]
657	4-Penten-1-ol	$C_5H_{10}O$	321.2	[71]
658	4-Penten-2-ol	$C_5H_{10}O$	303.2	[71]
659	2,2-Bis(chloromethyl)-1-propanol	$C_5H_{10}Cl_2O$	377.5	[71]
660	5-Hydroxy-2-pentanone (or 3-Acetyl-1-propanol)	$C_5H_{10}O_2$	366.5	[71]
661	3-Methyl-3-oxetanemethanol	$C_5H_{10}O_2$	372.0	[71]
662	Tetrahydrofurfuryl alcohol	$C_5H_{10}O_2$	357.0	[71]
663	Tetrahydro-4H-pyran-4-ol	$C_5H_{10}O_2$	360.9	[71]
664	3-Chloro-2,2-dimethyl-1-propanol	$C_5H_{11}ClO$	344.3	[71]
665	1-Methyl-3-pyrrolidinol	$C_5H_{11}NO$	343.7	[71]
666	2-Pyrrolidinemethanol	$C_5H_{11}NO$	359.3	[71]
667	3-Nitro-2-pentanol	$C_5H_{11}NO_3$	363.7	[71]
668	1-Pentanol	$C_5H_{12}O$	322.2	[257]
669	2-Pentanol	$C_5H_{12}O$	315.2	[71]
670	3-Pentanol	$C_5H_{12}O$	314.2	[255]
671	2-Methyl-1-butanol	$C_5H_{12}O$	329.2	[71]
672	3-Methyl-2-butanol	$C_5H_{12}O$	312.6	[256]
673	3-Methyl-1-butanol	$C_5H_{12}O$	316.2	[255]
674	2-Methyl-2-butanol (or tert-Amyl alcohol)	$C_5H_{12}O$	317.2	[71]
675	2,2-Dimethyl-1-propanol (or Neopentyl alcohol)	$C_5H_{12}O$	310.2	[165]
676	2,2-Dimethyl-1,3-propanediol (or Neopentyl glycol)	$C_5H_{12}O_2$	424.8	[71]
677	1,5-Pentanediol	$C_5H_{12}O_2$	408.2	[71]
678	1,2-Pentanediol	$C_5H_{12}O_2$	377.6	[71]
679	2,4-Pentanediol	$C_5H_{12}O_2$	374.8	[71]
680	2-Propoxyethanol	$C_5H_{12}O_2$	322.0	[71]
681	3-Ethoxy-1-propanol	$C_5H_{12}O_2$	327.6	[71]
682	1-Ethoxy-2-propanol	$C_5H_{12}O_2$	316.5	[71]
683	2-Isopropoxyethanol	$C_5H_{12}O_2$	316.2	[71]
684	3-Methoxy-1-butanol	$C_5H_{12}O_2$	347.2	[71]
685	2-(2-Methoxyethoxy)ethanol	$C_5H_{12}O_3$	369.3	[256]
686	3-Ethoxy-1,2-propanediol	$C_5H_{12}O_3$	383.2	[71]
687	2,2-bis(Hydroxymethyl)-1,3-propanediol (or Pentaerythritol)	$C_5H_{12}O_4$	468.2	[71]
688	2-Amino-3-methyl-1-butanol	$C_5H_{13}NO$	364.3	[71]
689	2-Amino-1-pentanol	$C_5H_{13}NO$	368.2	[71]
690	5-Amino-1-pentanol	$C_5H_{13}NO$	373.2	[71]
691	1-(Dimethylamino)-2-propanol	$C_5H_{13}NO$	314.2	[71]
692	2-(Dimethylamino)-2-propanol	$C_5H_{13}NO$	308.2	[71]
693	3-(Dimethylamino)-1-propanol	$C_5H_{13}NO$	314.2	[71]

(continued)

No.	Compound	Formula	Exp.	Ref.
694	2-(Isopropylamino) ethanol	$C_5H_{13}NO$	351.2	[71]
695	2-(Propylamino) ethanol	$C_5H_{13}NO$	360.9	[71]
696	N-Methyl diethanolamine	$C_5H_{13}NO_2$	399.8	[256]
697	3-(Dimethylamino)-1,2-propanediol	$C_5H_{13}NO_2$	377.6	[71]
698	2-((3-Aminopropyl)amino) ethanol	$C_5H_{14}N_2O$	425.2	[71]
699	2,3,4,5,6-Pentafluorophenol	C_6HF_5O	345.4	[165]
700	2,3,5,6-Tetrafluorophenol	$C_6H_2F_4O$	352.6	[71]
701	2,3,6-Trichlorophenol	$C_6H_3Cl_3O$	352.6	[71]
702	2,6-Dichloro-4-fluorophenol	$C_6H_3Cl_2FO$	352.6	[71]
703	3-Chloro-4-fluorophenol	C_6H_4ClFO	352.6	[71]
704	2-Bromo-4-fluorophenol	C_6H_4BrFO	352.6	[71]
705	2,4-Dibromophenol	$C_6H_4Br_2O$	383.2	[71]
706	5-Fluoro-2-nitrophenol	$C_6H_4FNO_3$	364.8	[71]
707	2,3-Difluorophenol	$C_6H_4F_2O$	329.8	[71]
708	2,4-Difluorophenol	$C_6H_4F_2O$	329.8	[71]
709	2,6-Difluorophenol	$C_6H_4F_2O$	332.0	[71]
710	2,4-Dichlorophenol	$C_6H_4Cl_2O$	387.0	[71]
711	2-Bromophenol	C_6H_5BrO	362.2	[71]
712	2-Chlorophenol	C_6H_5ClO	358.2	[71]
713	3-Chlorophenol	C_6H_5ClO	364.0	[165]
714	4-Chlorophenol	C_6H_5ClO	394.2	[71]
715	2-Fluorophenol	C_6H_5FO	319.8	[71]
716	3-Fluorophenol	C_6H_5FO	344.3	[71]
717	4-Fluorophenol	C_6H_5FO	341.5	[71]
718	3-Iodophenol	C_6H_5IO	383.2	[71]
719	2-Amino-4,6-dinitrophenol	$C_6H_5N_3O_5$	483.2	[71]
720	Phenol	C_6H_6O	358.2	[71]
721	1,2-Benzenediol (or Pyrocatechol or Catechol)	$C_6H_6O_2$	410.4	[71]
722	1,3-Benzendiol (or Resorcinol)	$C_6H_6O_2$	444.3	[71]
723	1,2,3-Benzenetriol	$C_6H_6O_3$	458.0	[165]
724	3-Methyl-1-penten-4-yn-3-ol	C_6H_8O	301.2	[71]
725	(trans) 3-Methyl-2-penten-4-yn-1-ol	C_6H_8O	332.2	[71]
726	(cis) 3-Methyl-2-penten-4-yn-1-ol	C_6H_8O	349.2	[71]
727	2-(Thiophen-2-yl)ethan-1-ol	C_6H_8OS	374.3	[71]
728	2-(3-Thienyl)ethanol	C_6H_8OS	365.9	[71]
729	2-Cyclohexen-1-ol	$C_6H_{10}O$	331.5	[71]
730	1,5-Hexadien-3-ol	$C_6H_{10}O$	302.6	[71]
731	2,4-Hexadien-1-ol	$C_6H_{10}O$	345.4	[71]
732	3,4-Dihydro-2H-pyran-2-methanol	$C_6H_{10}O_2$	362.0	[71]
733	2-Chlorocyclohexanol	$C_6H_{11}ClO$	343.2	[71]
734	Cyclohexanol	$C_6H_{12}O$	340.9	[71]
735	1-Hexen-3-ol	$C_6H_{12}O$	308.2	[71]
736	(cis) 2-Hexen-1-ol	$C_6H_{12}O$	334.8	[71]
737	(trans) 2-Hexen-1-ol	$C_6H_{12}O$	337.2	[71]

(continued)

(continued)

No.	Compound	Formula	Exp.	Ref.
738	(cis) 3-Hexen-1-ol	$C_6H_{12}O$	327.6	[71]
739	(trans) 3-Hexen-1-ol	$C_6H_{12}O$	332.0	[71]
740	4-Hexen-1-ol	$C_6H_{12}O$	334.3	[71]
741	4-Hexen-3-ol	$C_6H_{12}O$	319.2	[71]
742	5-Hexen-1-ol	$C_6H_{12}O$	320.4	[71]
743	1-Methylcyclopentanol	$C_6H_{12}O$	313.7	[71]
744	(trans) 2-Methylcyclopentanol	$C_6H_{12}O$	319.3	[71]
745	3-Methylcyclopentanol	$C_6H_{12}O$	328.2	[71]
746	3-Methyl-1-penten-3-ol	$C_6H_{12}O$	298.7	[71]
747	Cyclopentylmethanol (or Cyclopentanemethanol)	$C_6H_{12}O$	335.4	[71]
748	1,4-Cyclohexanediol	$C_6H_{12}O_2$	338.7	[71]
749	Tetrahydropyran-2-methanol	$C_6H_{12}O_2$	366.5	[71]
750	4-Hydroxy-4-methyl-2-pentanone (or Diacetone alcohol)	$C_6H_{12}O_2$	331.2	[255]
751	2,2-Dimethyl-1,3-dioxolane-4-methanol	$C_6H_{12}O_3$	363.2	[71]
752	6-Chloro-1-hexanol	$C_6H_{13}ClO$	372.0	[71]
753	2-[2-(2-Chloroethoxy)ethoxy]ethanol	$C_6H_{13}ClO_3$	380.4	[71]
754	1-Ethyl-3-pyrrolidinol	$C_6H_{13}NO$	347.0	[71]
755	(1-Aminocyclopentyl)methanol	$C_6H_{13}NO$	368.7	[71]
756	1-Methyl-2-pyrrolidinemethanol	$C_6H_{13}NO$	336.5	[71]
757	1-Hexanol	$C_6H_{14}O$	347.2	[71]
758	2-Hexanol	$C_6H_{14}O$	331.2	[71]
759	3-Hexanol	$C_6H_{14}O$	328.0	[71]
760	4-Methyl-1-pentanol	$C_6H_{14}O$	330.2	[71]
761	3-Methyl-2-pentanol	$C_6H_{14}O$	313.7	[71]
762	2-Methyl-3-pentanol	$C_6H_{14}O$	319.3	[71]
763	2-Methyl-2-pentanol	$C_6H_{14}O$	303.2	[71]
764	2-Methyl-1-pentanol	$C_6H_{14}O$	327.2	[255]
765	3-Methyl-1-pentanol	$C_6H_{14}O$	333.2	[71]
766	3-Methyl-3-pentanol	$C_6H_{14}O$	319.3	[71]
767	4-Methyl-2-pentanol	$C_6H_{14}O$	328.2	[71]
768	2,3-Dimethyl-2-butanol	$C_6H_{14}O$	302.6	[71]
769	3,3-Dimethyl-1-butanol	$C_6H_{14}O$	320.9	[71]
770	3,3-Dimethyl-2-butanol	$C_6H_{14}O$	302.0	[71]
771	2-Ethyl-1-butanol	$C_6H_{14}O$	336.2	[71]
772	1,6-Hexanediol (or Hexylene glycol)	$C_6H_{14}O_2$	403.2	[71]
773	2,5-Hexanediol	$C_6H_{14}O_2$	383.2	[71]
774	2-Methyl-1,3-pentanediol	$C_6H_{14}O_2$	394.0	[165]
775	2-Methyl-2,4-pentanediol	$C_6H_{14}O_2$	374.8	[71]
776	2,3-Dimethyl-2,3-butanediol (or Pinacol)	$C_6H_{14}O_2$	350.4	[71]
777	3,3-Dimethyl-1,2-butanediol	$C_6H_{14}O_2$	372.0	[71]
778	2-Butoxyethanol	$C_6H_{14}O_2$	347.0	[71]
779	2-Isobutoxyethanol	$C_6H_{14}O_2$	330.9	[71]

(continued)

No.	Compound	Formula	Exp.	Ref.
780	1-Propoxy-2-propanol	$C_6H_{14}O_2$	321.2	[165]
781	Dipropylene glycol	$C_6H_{14}O_3$	410.9	[71]
782	2-(2-Ethoxyethoxy)ethanol	$C_6H_{14}O_3$	372.0	[71]
783	2-Ethyl-2-(hydroxymethyl)-1,3-propanediol	$C_6H_{14}O_3$	452.6	[71]
784	Triethylene glycol	$C_6H_{14}O_4$	450.2	[255]
785	6-Aminohexanol	$C_6H_{15}NO$	390.0	[165]
786	2-Amino-1-hexanol	$C_6H_{15}NO$	373.2	[71]
787	2-(Butylamino)ethanol	$C_6H_{15}NO$	360.2	[71]
788	2-(tert-Butylamino)ethanol	$C_6H_{15}NO$	360.2	[71]
789	2-(Dimethylamino)-2-methyl-1-propanol	$C_6H_{15}NO$	338.7	[71]
790	2-(Diethylamino)ethanol	$C_6H_{15}NO$	333.2	[71]
791	Isoleucinol	$C_6H_{15}NO$	373.7	[71]
792	Diisopropanolamine	$C_6H_{15}NO_2$	399.8	[71]
793	N-Ethyldiethanolamine	$C_6H_{15}NO_2$	410.9	[71]
794	Triethanolamine	$C_6H_{15}NO_3$	469.3	[71]
795	(Perfluorophenyl)methanol	$C_7H_3F_5O$	360.9	[71]
796	2-Nitro-4-(trifluoromethyl)phenol	$C_7H_4F_3NO_3$	368.2	[71]
797	(2,3-Difluorophenyl)methanol	$C_7H_6F_2O$	367.0	[71]
798	(2,4-Difluorophenyl)methanol	$C_7H_6F_2O$	363.2	[71]
799	(2,5-Difluorophenyl)methanol	$C_7H_6F_2O$	367.0	[71]
800	(2,6-Difluorophenyl)methanol	$C_7H_6F_2O$	362.0	[71]
801	(3,4-Difluorophenyl)methanol	$C_7H_6F_2O$	370.9	[71]
802	(3,5-Difluorophenyl)methanol	$C_7H_6F_2O$	367.6	[71]
803	(2-Fluorophenyl)methanol	C_7H_7FO	363.7	[71]
804	(3-Fluorophenyl)methanol	C_7H_7FO	363.7	[71]
805	(4-Fluorophenyl)methanol	C_7H_7FO	363.2	[71]
806	2-Chloro-5-methylphenol	C_7H_7ClO	354.3	[71]
807	3-Methyl-2-nitrophenol	$C_7H_7NO_3$	380.4	[71]
808	4-Methyl-2-nitrophenol	$C_7H_7NO_3$	381.5	[71]
809	5-Methyl-2-nitrophenol	$C_7H_7NO_3$	382.6	[71]
810	Phenylmethanol (or Benzyl alcohol)	C_7H_8O	377.6	[71]
811	2-Methylphenol (or o-Cresol)	C_7H_8O	354.2	[255]
812	3-Methylphenol (or m-Cresol)	C_7H_8O	359.2	[255]
813	4-Methylphenol (or p-Cresol)	C_7H_8O	359.2	[255]
814	2-(Methylthio)phenol	C_7H_8OS	368.7	[71]
815	2-Methoxyphenol (or Guaiacol)	$C_7H_8O_2$	355.4	[71]
816	4-Methoxyphenol	$C_7H_8O_2$	405.4	[71]
817	1-Ethynylcyclopentanol	$C_7H_{10}O$	322.0	[71]
818	3-Methyl-2-cyclohexen-1-ol	$C_7H_{12}O$	344.8	[71]
819	3-Cyclohexene-1-methanol	$C_7H_{12}O$	349.3	[71]
820	3-Ethyl-1-pentyn-3-ol	$C_7H_{12}O$	310.2	[71]
821	1,6-Heptadien-4-ol	$C_7H_{12}O$	313.2	[71]
822	Cycloheptanol	$C_7H_{14}O$	344.3	[71]
823	Cyclohexylmethanol	$C_7H_{14}O$	350.2	[71]

(continued)

(continued)

No.	Compound	Formula	Exp.	Ref.
824	1-Hepten-3-ol	$C_7H_{14}O$	327.2	[71]
825	1-Methylcyclohexanol	$C_7H_{14}O$	340.9	[71]
826	(cis) 2-Methylcyclohexanol	$C_7H_{14}O$	332.0	[71]
827	(trans) 2-Methylcyclohexanol	$C_7H_{14}O$	332.0	[71]
828	(cis) 3-Methylcyclohexanol	$C_7H_{14}O$	335.2	[165]
829	(trans) 3-Methylcyclohexanol	$C_7H_{14}O$	343.2	[71]
830	(cis) 4-Methylcyclohexanol	$C_7H_{14}O$	343.2	[165]
831	(trans) 4-Methylcyclohexanol	$C_7H_{14}O$	343.2	[165]
832	1-Methyl-2-piperidinemethanol	$C_7H_{15}NO$	354.3	[71]
833	1-Methyl-3-piperidinemethanol	$C_7H_{15}NO$	367.6	[71]
834	1-Piperidineethanol	$C_7H_{15}NO$	342.0	[71]
835	2-Piperidineethanol	$C_7H_{15}NO$	375.4	[71]
836	1-Methyl-2-pyrrolidineethanol	$C_7H_{15}NO$	357.6	[71]
837	1-Heptanol	$C_7H_{16}O$	349.8	[256]
838	2-Heptanol	$C_7H_{16}O$	344.3	[71]
839	3-Heptanol	$C_7H_{16}O$	333.2	[71]
840	4-Heptanol	$C_7H_{16}O$	327.2	[71]
841	2-Methyl-1-hexanol	$C_7H_{16}O$	333.0	[165]
842	2-Methyl-2-hexanol	$C_7H_{16}O$	313.7	[71]
843	2-Methyl-3-hexanol	$C_7H_{16}O$	313.7	[71]
844	5-Methyl-1-hexanol	$C_7H_{16}O$	340.0	[165]
845	5-Methyl-2-hexanol	$C_7H_{16}O$	319.3	[71]
846	3-Ethyl-3-pentanol	$C_7H_{16}O$	313.2	[71]
847	2,2-Dimethyl-3-pentanol	$C_7H_{16}O$	310.9	[71]
848	2,3-Dimethyl-3-pentanol	$C_7H_{16}O$	313.7	[71]
849	2,4-Dimethyl-3-pentanol	$C_7H_{16}O$	323.2	[71]
850	4,4-Dimethyl-2-pentanol	$C_7H_{16}O$	309.8	[71]
851	3-(Methylthio)-1-hexanol	$C_7H_{16}OS$	380.9	[71]
852	1-Butoxy-2-propanol	$C_7H_{16}O_2$	335.2	[165]
853	2,2-Diethyl-1,3-propanediol	$C_7H_{16}O_2$	374.8	[71]
854	2,4-Dimethyl-2,4-pentanediol	$C_7H_{16}O_2$	369.8	[71]
855	1-Diethylamino-2-propanol	$C_7H_{17}NO$	306.5	[71]
856	3-Diethylamino-1-propanol	$C_7H_{17}NO$	338.7	[71]
857	3-Diethylamino-1,2-propanediol	$C_7H_{17}NO_2$	380.9	[71]
858	2-((2-(Dimethylamino)ethyl)(methyl)amino) ethan-1-ol	$C_7H_{18}N_2O$	359.8	[71]
859	1-(Pentafluorophenyl)ethanol	$C_8H_5F_5O$	360.4	[71]
860	(2-(Trifluoromethyl)phenyl)methanol	$C_8H_7F_3O$	365.4	[71]
861	(3-(Trifluoromethyl)phenyl)methanol	$C_8H_7F_3O$	357.6	[71]
862	(4-(Trifluoromethyl)phenyl)methanol	$C_8H_7F_3O$	373.7	[71]
863	1-(2,4-Dichlorophenyl)ethanol	$C_8H_8Cl_2O$	382.0	[71]
864	4-Vinylphenol (or 4-Hydroxystyrene)	C_8H_8O	377.0	[165]
865	1-(4-Bromophenyl)ethanol	C_8H_9BrO	336.5	[71]
866	2-(2-Fluorophenyl)ethanol	C_8H_9FO	371.5	[71]

(continued)

No.	Compound	Formula	Exp.	Ref.
867	2-(4-Fluorophenyl)ethanol	C_8H_9FO	377.6	[71]
868	2-Ethylphenol	$C_8H_{10}O$	351.5	[71]
869	3-Ethylphenol	$C_8H_{10}O$	367.2	[165]
870	4-Ethylphenol	$C_8H_{10}O$	377.0	[71]
871	2-Methylbenzyl alcohol (o-Tolylmethanol)	$C_8H_{10}O$	377.6	[71]
872	3-Methylbenzyl alcohol (m-Tolylmethanol)	$C_8H_{10}O$	378.7	[71]
873	1-Phenylethanol	$C_8H_{10}O$	369.3	[71]
874	2-Phenylethanol (or Benzeneethanol)	$C_8H_{10}O$	375.4	[71]
875	2,3-Dimethylphenol (or 2,3-Xylenol)	$C_8H_{10}O$	365.0	[165]
876	2,4-Dimethylphenol (or 2,4-Xylenol)	$C_8H_{10}O$	368.2	[165]
877	2,5-Dimethylphenol (or 2,5-Xylenol)	$C_8H_{10}O$	360.0	[165]
878	2,6-Dimethylphenol (or 2,6-Xylenol)	$C_8H_{10}O$	360.9	[71]
879	3,4-Dimethylphenol (or 3,4-Xylenol)	$C_8H_{10}O$	376.0	[165]
880	2-(Ethylthio)phenol	$C_8H_{10}OS$	366.5	[71]
881	2-Ethoxyphenol	$C_8H_{10}O_2$	364.3	[71]
882	2-Methoxy-4-methylphenol	$C_8H_{10}O_2$	372.6	[71]
883	2-Phenoxyethanol	$C_8H_{10}O_2$	399.8	[71]
884	2,3-Dimethoxyphenol	$C_8H_{10}O_3$	382.0	[71]
885	3,5-Dimethoxyphenol	$C_8H_{10}O_3$	351.5	[71]
886	1-(3-Aminophenyl)ethanol	$C_8H_{11}NO$	430.4	[71]
887	2-(Ahenylamino)ethanol (or 2-Anilinoethanol)	$C_8H_{11}NO$	425.2	[71]
888	l-Ethynyl-1-cyclohexanol	$C_8H_{12}O$	335.9	[71]
889	Bicyclo[2.2.1]hept-5-en-2-ylmethanol	$C_8H_{12}O$	359.8	[71]
890	3,5-Dimethyl-1-hexyn-3-ol	$C_8H_{14}O$	330.4	[71]
891	Bicyclo[2.2.1]heptanylmethanol	$C_8H_{14}O$	357.6	[71]
892	1-Octyn-3-ol	$C_8H_{14}O$	339.2	[71]
893	2,3-Dimethylcyclohexanol	$C_8H_{16}O$	338.7	[71]
894	2,5-Dimethylcyclohexanol	$C_8H_{16}O$	343.2	[71]
895	2,6-Dimethylcyclohexanol	$C_8H_{16}O$	343.2	[71]
896	3,4-Dimethylcyclohexanol	$C_8H_{16}O$	351.5	[71]
897	3,5-Dimethylcyclohexanol	$C_8H_{16}O$	346.5	[71]
898	2-Ethylcyclohexanol	$C_8H_{16}O$	341.5	[71]
899	4-Ethylcyclohexanol	$C_8H_{16}O$	350.9	[71]
900	3-Cyclopentyl-1-propanol	$C_8H_{16}O$	355.4	[71]
901	2-Cyclohexylethanol	$C_8H_{16}O$	359.2	[71]
902	(4-Methylcyclohexyl)methanol	$C_8H_{16}O$	354.0	[165]
903	1-Cyclohexylethanol	$C_8H_{16}O$	345.9	[71]
904	Cycloheptylmethanol	$C_8H_{16}O$	365.9	[71]
905	Cyclooctanol	$C_8H_{16}O$	361.2	[71]
906	1-Octen-3-ol	$C_8H_{16}O$	360.2	[71]
907	6-Methyl-5-hepten-2-ol	$C_8H_{16}O$	340.0	[71]
908	1,4-Cyclohexanedimethanol	$C_8H_{16}O_2$	440.2	[71]
909	4-[(Tetrahydro-2-furanyl)oxy]-1-butanol	$C_8H_{16}O_3$	402.0	[165]
910	1-Octanol	$C_8H_{18}O$	363.7	[71]

(continued)

(continued)

No.	Compound	Formula	Exp.	Ref.
911	2-Octanol	$C_8H_{18}O$	361.2	[255]
912	3-Octanol	$C_8H_{18}O$	341.2	[71]
913	4-Octanol	$C_8H_{18}O$	338.2	[71]
914	2-Ethyl-1-hexanol	$C_8H_{18}O$	358.2	[71]
915	2-Ethyl-4-methyl-1-pentanol	$C_8H_{18}O$	350.2	[71]
916	6-Methyl-1-heptanol (or Isooctyl alcohol)	$C_8H_{18}O$	355.4	[71]
917	2,4,4-Trimethyl-1-pentanol	$C_8H_{18}O$	333.2	[71]
918	6-Methyl-2-heptanol	$C_8H_{18}O$	340.4	[71]
919	4-Methyl-3-heptanol	$C_8H_{18}O$	327.6	[71]
920	2-Propyl-1-pentanol	$C_8H_{18}O$	344.8	[71]
921	4,5-Octanediol	$C_8H_{18}O_2$	383.2	[71]
922	2-Ethyl-1,3-hexanediol	$C_8H_{18}O_2$	402.6	[71]
923	2,2,4-Trimethyl-1,3-pentanediol	$C_8H_{18}O_2$	385.9	[71]
924	2-(2-Ethylbutoxy)ethanol	$C_8H_{18}O_2$	355.4	[71]
925	2-(Hexyloxy)-1-ethanol	$C_8H_{18}O_2$	363.7	[71]
926	2-(2-Butoxyethoxy)ethanol	$C_8H_{18}O_3$	388.8	[71]
927	Tetraethylene glycol	$C_8H_{18}O_5$	469.3	[256]
928	2-(Diisopropylamino)ethanol	$C_8H_{19}NO$	352.6	[71]
929	2-(2-(Diethylamino)ethoxy)ethanol	$C_8H_{19}NO_2$	369.2	[71]
930	2,2'-(Butylazanediyl)bis(ethanol)	$C_8H_{19}NO_2$	419.3	[71]
931	2,2'-(tert-Butylazanediyl)bis(ethanol)	$C_8H_{19}NO_2$	413.7	[71]
932	(3,5-Bis(trifluoromethyl)phenyl)methanol	$C_9H_6F_6O$	370.9	[71]
933	1-(2-(Trifluoromethyl)phenyl)ethanol	$C_9H_9F_3O$	362.6	[71]
934	2-(2-(Trifluoromethyl)phenyl)ethanol	$C_9H_9F_3O$	375.9	[71]
935	2-(3-(Trifluoromethyl)phenyl)ethanol	$C_9H_9F_3O$	358.2	[71]
936	4-(1-Propen-2-yl)phenol	$C_9H_{10}O$	385.0	[165]
937	2-Allylphenol	$C_9H_{10}O$	362.0	[71]
938	1-Phenyl-1-propanol	$C_9H_{12}O$	363.2	[165]
939	1-Phenyl-2-propanol	$C_9H_{12}O$	358.2	[165]
940	2-Phenyl-1-propanol	$C_9H_{12}O$	366.2	[165]
941	2-Phenyl-2-propanol	$C_9H_{12}O$	360.2	[165]
942	3-Phenyl-1-propanol	$C_9H_{12}O$	382.6	[71]
943	(2,5-Dimethylphenyl)methanol	$C_9H_{12}O$	380.4	[71]
944	(3,5-Dimethylphenyl)methanol	$C_9H_{12}O$	379.8	[71]
945	(4-Ethylphenyl)methanol	$C_9H_{12}O$	377.6	[71]
946	2-Isopropylphenol	$C_9H_{12}O$	380.9	[71]
947	3-Isopropylphenol	$C_9H_{12}O$	370.4	[71]
948	2-(m-Tolyl)ethanol	$C_9H_{12}O$	382.6	[71]
949	2-(p-Tolyl)ethanol	$C_9H_{12}O$	380.4	[71]
950	2-Propylphenol	$C_9H_{12}O$	366.5	[71]
951	4-Propylphenol	$C_9H_{12}O$	379.3	[71]
952	2-(Benzyloxy)ethanol	$C_9H_{12}O_2$	402.2	[71]
953	2-Isopropoxyphenol	$C_9H_{12}O_2$	360.9	[71]
954	2-Methoxy-2-phenylethanol	$C_9H_{12}O_2$	370.9	[71]

(continued)

No.	Compound	Formula	Exp.	Ref.
955	3-Phenoxy-1,2-propanediol	$C_9H_{12}O_3$	429.6	[71]
956	2-(Methyl(phenyl)amino) ethanol	$C_9H_{13}NO$	410.9	[71]
957	6-Methyl-2-pyridinepropanol	$C_9H_{13}NO$	383.2	[71]
958	2-(p-Tolylamino)ethanol	$C_9H_{13}NO$	416.5	[71]
959	2,4-Dimethyl-2,6-heptadien-1-ol	$C_9H_{16}O$	351.5	[71]
960	3-Methyl-2-norbornanemethanol	$C_9H_{16}O$	363.7	[71]
961	3-Nonyn-1-ol	$C_9H_{16}O$	367.0	[71]
962	3,5,5-Trimethyl-2-cyclohexen-1-ol	$C_9H_{16}O$	353.2	[71]
963	Cyclooctanemethanol	$C_9H_{18}O$	366.5	[71]
964	3,3,5-Trimethyl-1-cyclohexanol	$C_9H_{18}O$	360.9	[71]
965	3-Nonen-1-ol	$C_9H_{18}O$	335.4	[71]
966	1-Nonanol	$C_9H_{20}O$	369.2	[257]
967	2-Nonanol	$C_9H_{20}O$	369.2	[71]
968	3-Nonanol	$C_9H_{20}O$	369.2	[71]
969	5-Nonanol	$C_9H_{20}O$	350.2	[71]
970	4-Methyl-1-octanol	$C_9H_{20}O$	358.0	[165]
971	6-Methyl-1-octanol	$C_9H_{20}O$	360.0	[165]
972	2,6-Dimethyl-4-heptanol	$C_9H_{20}O$	347.0	[71]
973	3-Ethyl-1-heptanol	$C_9H_{20}O$	363.0	[165]
974	3,5,5-Trimethyl-1-hexanol	$C_9H_{20}O$	366.5	[71]
975	2-(2-Pentoxyethoxy)ethanol	$C_9H_{20}O_3$	383.0	[165]
976	1-(2-Butoxyethoxy)-2-propanol	$C_9H_{20}O_3$	394.3	[71]
977	Tripropylene glycol	$C_9H_{20}O_4$	413.7	[71]
978	3-(Dipropylamino)-1,2-propanediol	$C_9H_{21}NO_2$	383.2	[71]
979	3-(Diisopropylamino)-1,2-propanediol	$C_9H_{21}NO_2$	383.2	[71]
980	1,1′,1″-Nitrilotris(2-propanol) (or Triisopropanolamine)	$C_9H_{21}NO_3$	433.2	[71]
981	2-Naphthalenol (or 2-Naphthol)	$C_{10}H_8O$	434.2	[71]
982	2-Phenyl-3-butyn-2-ol	$C_{10}H_{10}O$	369.3	[71]
983	2-Allyl-4-methylphenol	$C_{10}H_{12}O$	374.8	[71]
984	2-Allyl-6-methylphenol	$C_{10}H_{12}O$	367.6	[71]
985	Cyclopropyl(phenyl)methanol	$C_{10}H_{12}O$	385.4	[71]
986	2-(sec-Butyl)-4,6-dinitrophenol	$C_{10}H_{12}N_2O_5$	450.2	[165]
987	4-(tert-Butyl)-2-chlorophenol	$C_{10}H_{13}ClO$	380.4	[71]
988	3-Isopropyl-5-methylphenol	$C_{10}H_{14}O$	383.2	[165]
989	2-Isopropyl-5-methylphenol (or Thymol)	$C_{10}H_{14}O$	375.4	[71]
990	5-Isopropyl-2-methylphenol (or Carvacrol)	$C_{10}H_{14}O$	379.8	[71]
991	2-(sec-Butyl)phenol	$C_{10}H_{14}O$	385.4	[71]
992	2-(tert-Butyl)phenol	$C_{10}H_{14}O$	383.2	[71]
993	3-(tert-Butyl)phenol	$C_{10}H_{14}O$	382.0	[71]
994	4-(sec-Butyl)phenol	$C_{10}H_{14}O$	388.7	[71]
995	1-Phenyl-2-butanol (or alpha-Ethylphenethyl alcohol)	$C_{10}H_{14}O$	373.2	[71]
996	2-Methyl-1-phenyl-1-propanol	$C_{10}H_{14}O$	359.8	[71]

(continued)

(continued)

No.	Compound	Formula	Exp.	Ref.
997	2-Methyl-1-phenyl-2-propanol	$C_{10}H_{14}O$	354.3	[71]
998	2-Phenyl-2-butanol	$C_{10}H_{14}O$	363.7	[71]
999	4-(tert-Butyl)-1,2-benzenediol (or p-tert-Butylcatechol)	$C_{10}H_{14}O_2$	424.8	[71]
1000	(2,4-Dimethoxy-3-methylphenyl)methanol	$C_{10}H_{14}O_3$	383.2	[71]
1001	(2,3,4-Trimethoxyphenyl)methanol	$C_{10}H_{14}O_4$	407.6	[71]
1002	2-(Ethyl(phenyl)amino)-1-ethanol	$C_{10}H_{15}NO$	405.4	[71]
1003	Carveol	$C_{10}H_{16}O$	371.5	[71]
1004	Myrtenol	$C_{10}H_{16}O$	362.6	[71]
1005	4-Ethyl-1-octyn-3-ol	$C_{10}H_{18}O$	356.2	[71]
1006	Borneol	$C_{10}H_{18}O$	338.7	[71]
1007	Dihydrocarveol	$C_{10}H_{18}O$	364.8	[71]
1008	para-Menth-1-en-9-ol	$C_{10}H_{18}O$	376.5	[71]
1009	Fenchyl alcohol	$C_{10}H_{18}O$	347.0	[71]
1010	Geraniol	$C_{10}H_{18}O$	349.8	[71]
1011	Isopinocampheol	$C_{10}H_{18}O$	366.5	[71]
1012	5-Methyl-2-(1-propen-2-yl)-1-cyclohexanol (or Isopulegol)	$C_{10}H_{18}O$	351.5	[71]
1013	3,7-Dimethyl-1,6-octadien-3-ol (or Linalool)	$C_{10}H_{18}O$	349.3	[71]
1014	(cis) Myrtanol	$C_{10}H_{18}O$	341.2	[71]
1015	(trans) Myrtanol	$C_{10}H_{18}O$	377.0	[71]
1016	Nerol	$C_{10}H_{18}O$	373.2	[71]
1017	2-(tert-Butyl)cyclohexanol	$C_{10}H_{20}O$	352.6	[71]
1018	4-(tert-Butyl)cyclohexanol	$C_{10}H_{20}O$	378.2	[71]
1019	3,7-Dimethyl-6-octen-1-ol (or beta-Citronellol)	$C_{10}H_{20}O$	372.2	[71]
1020	(trans) 5-Decen-1-ol	$C_{10}H_{20}O$	335.4	[71]
1021	9-Decen-1-ol	$C_{10}H_{20}O$	372.0	[71]
1022	4-Cyclohexyl-1-butanol	$C_{10}H_{20}O$	382.0	[71]
1023	2,6-Dimethyl-7-octen-2-ol (or Dihydromyrcenol)	$C_{10}H_{20}O$	349.8	[71]
1024	2-Isopropyl-5-methylcyclohexanol (or Menthol)	$C_{10}H_{20}O$	366.5	[71]
1025	1-Decanol	$C_{10}H_{22}O$	385.9	[71]
1026	2-Decanol	$C_{10}H_{22}O$	358.2	[71]
1027	4-Decanol	$C_{10}H_{22}O$	355.4	[71]
1028	8-Methyl-1-nonanol (or Isodecanol)	$C_{10}H_{22}O$	377.6	[71]
1029	3,7-Dimethyl-1-octanol	$C_{10}H_{22}O$	370.2	[71]
1030	2-(2-Hexoxyethoxy)ethanol	$C_{10}H_{22}O_3$	408.2	[165]
1031	Dipropylene glycol butyl ether	$C_{10}H_{22}O_3$	386.2	[71]
1032	Triethylene glycol monobutyl ether	$C_{10}H_{22}O_4$	416.5	[71]
1033	Tripropylene glycol, monomethyl ether	$C_{10}H_{22}O_4$	399.8	[71]
1034	2-(Dibutylamino) ethanol	$C_{10}H_{23}NO$	377.6	[71]
1035	1 Benzyl-3-pyrrolidinol	$C_{11}H_{15}NO$	383.2	[71]
1036	2-Pentylphenol	$C_{11}H_{16}O$	377.2	[71]

(continued)

No.	Compound	Formula	Exp.	Ref.
1037	4-Pentylphenol	$C_{11}H_{16}O$	385.2	[71]
1038	2-(Pentan-2-yl)phenol	$C_{11}H_{16}O$	366.5	[71]
1039	4-(Pentan-2-yl)phenol	$C_{11}H_{16}O$	405.4	[71]
1040	4-(tert-Pentyl)phenol	$C_{11}H_{16}O$	384.3	[71]
1041	2-(tert-Butyl)-4-methylphenol	$C_{11}H_{16}O$	373.2	[71]
1042	2-(tert-Butyl)-5-methylphenol	$C_{11}H_{16}O$	376.7	[71]
1043	2-(tert-Butyl)-6-methylphenol	$C_{11}H_{16}O$	380.4	[71]
1044	4-(tert-Butyl)-2-methylphenol	$C_{11}H_{16}O$	390.9	[71]
1045	2-(tert-Butyl)-3-methylphenol	$C_{11}H_{16}O$	377.8	[71]
1046	1-Phenyl-2-pentanol	$C_{11}H_{16}O$	378.2	[71]
1047	2,2-Dimethyl-3-phenyl-1-propanol	$C_{11}H_{16}O$	382.6	[71]
1048	2-(1-Phenylethoxy)-1-propanol	$C_{11}H_{16}O_2$	386.0	[165]
1049	10-Undecen-1-ol	$C_{11}H_{22}O$	366.5	[71]
1050	4-(tert-Pentyl)-1-cyclohexanol	$C_{11}H_{22}O$	373.1	[71]
1051	1-Undecanol	$C_{11}H_{24}O$	388.0	[165]
1052	2-Undecanol	$C_{11}H_{24}O$	385.9	[71]
1053	5-Ethyl-2-nonanol	$C_{11}H_{24}O$	373.2	[71]
1054	2,4-Diethyl-1-heptanol	$C_{11}H_{24}O$	373.7	[71]
1055	1-(Dibutylamino)-2-propanol	$C_{11}H_{25}NO$	369.2	[71]
1056	2-Chloro-4-phenylphenol	$C_{12}H_9ClO$	447.0	[71]
1057	2-Phenylphenol	$C_{12}H_{10}O$	397.0	[71]
1058	4-phenylphenol	$C_{12}H_{10}O$	438.7	[71]
1059	3-Phenoxyphenol	$C_{12}H_{10}O_2$	383.2	[71]
1060	2-Phenyl-1-cyclohexanol	$C_{12}H_{16}O$	410.9	[71]
1061	2-Cyclohexylphenol	$C_{12}H_{16}O$	407.0	[71]
1062	(1-Benzylpyrrolidin-2-yl)methanol	$C_{12}H_{17}NO$	383.2	[71]
1063	2-(tert-Butyl)-4,6-dimethylphenol	$C_{12}H_{18}O$	384.8	[71]
1064	5-Methyl-2-pentylphenol	$C_{12}H_{18}O$	388.9	[71]
1065	2-(4-(tert-Butyl)phenoxy)-1-ethanol	$C_{12}H_{18}O_2$	430.4	[71]
1066	Dipropylene glycol phenyl ether	$C_{12}H_{18}O_3$	430.4	[71]
1067	(cis) 7-Dodecen-1-ol	$C_{12}H_{24}O$	334.3	[71]
1068	1-Dodecanol	$C_{12}H_{26}O$	402.6	[71]
1069	2-Methyl-1-undecanol	$C_{12}H_{26}O$	392.0	[165]
1070	2-Butyl-1-octanol	$C_{12}H_{26}O$	383.2	[71]
1071	2,6,8-Trimethyl-4-nonanol	$C_{12}H_{26}O$	366.5	[71]
1072	2-(4-tert-Pentylphenoxy) ethanol	$C_{13}H_{20}O_2$	410.9	[71]
1073	1-(4-(sec-Butyl)phenoxy)-2-propanol	$C_{13}H_{20}O_2$	419.3	[71]
1074	2-Butyl-1-nonanol	$C_{13}H_{28}O$	404.0	[165]
1075	2-Methyl-dodecan-1-ol	$C_{13}H_{28}O$	404.0	[165]
1076	1-Tridecanol	$C_{13}H_{28}O$	394.3	[71]
1077	4-(2-Methyl-2-heptanyl)phenol	$C_{14}H_{22}O$	406.0	[165]
1078	2,4-Di-tert-butylphenol	$C_{14}H_{22}O$	402.6	[71]
1079	2,6-Di-tert-butylphenol	$C_{14}H_{22}O$	391.5	[71]
1080	(cis) 7-Tetradecen-1-ol	$C_{14}H_{28}O$	334.3	[71]

(continued)

(continued)

No.	Compound	Formula	Exp.	Ref.
1081	(*cis*) 9-Tetradecen-1-ol	$C_{14}H_{28}O$	335.4	[71]
1082	(*cis*) 11-Tetradecen-1-ol	$C_{14}H_{28}O$	335.4	[71]
1083	1-Tetradecanol	$C_{14}H_{30}O$	421.2	[257]
1084	2-Methyl-1-tridecanol	$C_{14}H_{30}O$	422.0	[165]
1085	7-Ethyl-2-methyl-4-undecanol	$C_{14}H_{30}O$	413.7	[71]
1086	4-(2-Phenylpropan-2-yl)phenol (or p-Cumylphenol)	$C_{15}H_{16}O$	433.0	[165]
1087	4,4′-(Propane-2,2-diyl)diphenol (or Bisphenol a)	$C_{15}H_{16}O_2$	486.0	[256]
1088	2,6-Di-tert-butyl-4-methylphenol	$C_{15}H_{24}O$	400.2	[71]
1089	2,4-Di-tert-butyl-5-methylphenol	$C_{15}H_{24}O$	400.9	[71]
1090	1-Pentadecanol	$C_{15}H_{32}O$	424.0	[165]
1091	4-(tert-Butyl)-2-phenylphenol	$C_{16}H_{18}O$	433.2	[71]
1092	2-Chloro-4,6-di-tert-pentylphenol	$C_{16}H_{25}ClO$	394.3	[71]
1093	2,4-Dipentylphenol	$C_{16}H_{26}O$	400.2	[71]
1094	11-Hexadecen-1-ol	$C_{16}H_{32}O$	334.3	[71]
1095	1-Hexadecanol	$C_{16}H_{34}O$	433.0	[165]
1096	1-Heptadecanol	$C_{17}H_{36}O$	441.0	[165]
1097	3,9-Diethyl-6-tridecanol	$C_{17}H_{36}O$	427.6	[71]
1098	1-Octadecanol	$C_{18}H_{38}O$	452.6	[71]
1099	1-Nonadecanol	$C_{19}H_{40}O$	458.0	[165]
1100	1-Icosanol	$C_{20}H_{42}O$	467.0	[165]
1101	Hydrocyanic acid	CHN	255	[43]
1102	Formic acid	CH_2O_2	321.15	[43]
1103	Acetic acid	$C_2H_4O_2$	316	[43]
1104	Acetic acid, dichloro-	$C_2H_2Cl_2O_2$	383.15	[43]
1105	Acetic acid, mercapto-	$C_2H_4O_2S$	398	[43]
1106	Carbonochloridic acid, methyl ester	$C_2H_3ClO_2$	285.15	[43]
1107	Ethaneperoxoic acid	$C_2H_4O_3$	313.15	[43]
1108	Sulfuric acid, dimethyl ester	$C_2H_6O_4S$	356.15	[43]
1109	Sulfuric acid, monoethyl ester	$C_2H_6O_4S$	457	[43]
1110	Formic acid, methyl ester	$C_2H_4O_2$	254	[43]
1111	2-Propenoic acid	$C_3H_4O_2$	324	[43]
1112	Acetic acid, chloro-, methyl ester	$C_3H_5ClO_2$	324.816	[43]
1113	Acetic acid, hydroxy-, methylester	$C_3H_6O_3$	339.15	[43]
1114	Acetic acid, methoxy-	$C_3H_6O_3$	388	[43]
1115	Acetic acid, methyl ester	$C_3H_6O_2$	263.15	[43]
1116	Carbonic acid, dimethyl ester	$C_3H_6O_3$	289.85	[43]
1117	Carbonochloridic acid, ethyl ester	$C_3H_5ClO_2$	275.15	[43]
1118	Formic acid, ethenyl ester	$C_3H_4O_2$	254	[43]
1119	Formic acid, ethyl ester	$C_3H_6O_2$	253.15	[43]
1120	Phosphoric acid, trimethyl ester	$C_3H_9O_4P$	363	[43]
1121	Propanedioic acid	$C_3H_4O_4$	496	[43]
1122	Propanoic acid	$C_3H_6O_2$	328	[43]

(continued)

No.	Compound	Formula	Exp.	Ref.
1123	(+–)-2-Hydroxy-propanoic acid	$C_3H_6O_3$	410	[43]
1124	2-Oxy- propanoic acid	$C_3H_4O_3$	355.15	[43]
1125	(E)- 2- Butene dioic acid	$C_4H_4O_4$	471	[43]
1126	(E)- 2-Butenoic acid	$C_4H_6O_2$	368	[43]
1127	(Z)- 2- Butenoic acid	$C_4H_6O_2$	352	[43]
1128	2-methyl-2-Propenoic acid	$C_4H_6O_2$	340	[43]
1129	2-Propenoic acid, methyl ester	$C_4H_6O_2$	270	[43]
1130	Acetic acid, (acetyloxy)	$C_4H_6O_4$	440	[43]
1131	2,2'-Oxybis- acetic acid	$C_4H_6O_5$	500	[43]
1132	Acetic acid, anhydride	$C_4H_6O_3$	322.594	[43]
1133	Acetic acid, chloro-, ethyl ester	$C_4H_7ClO_2$	311	[43]
1134	Acetic acid, cyano-, methyl ester	$C_4H_5NO_2$	383.15	[43]
1135	Acetic acid, ethenyl ester	$C_4H_6O_2$	265	[43]
1136	Acetic acid, ethyl ester	$C_4H_8O_2$	269	[43]
1137	Butanedioic acid	$C_4H_6O_4$	489	[43]
1138	Butanedioic acid, hydroxy-(+–)-	$C_4H_6O_5$	515	[43]
1139	Butanoic acid	$C_4H_8O_2$	345	[43]
1140	Cyclopropanecarboxylic acid	$C_4H_6O_2$	344.15	[43]
1141	Ethanethioic acid, S-ethyl ester	C_4H_8OS	290	[43]
1142	Formic acid, propyl ester	$C_4H_8O_2$	270	[43]
1143	2-Hydroxy-propanoic acid, methyl ester	$C_4H_8O_3$	322.15	[43]
1144	2-Hydroxy-2-methyl- propanoic acid	$C_4H_8O_3$	396	[43]
1145	2-Methyl- propanoic acid	$C_4H_8O_2$	329.15	[43]
1146	Propanoic acid, methyl ester	$C_4H_8O_2$	271	[43]
1147	Sulfuric acid, diethyl ester	$C_4H_{10}O_4S$	351.15	[43]
1148	Sulfurous acid, diethyl ester	$C_4H_{10}O_3S$	326.15	[43]
1149	Butanoic acid, methyl ester	$C_5H_{10}O_2$	287	[43]
1150	Carbonic acid, diethyl ester	$C_5H_{10}O_3$	298.15	[43]
1151	Methylene- butanedioic acid	$C_5H_6O_4$	488	[43]
1152	2-Methyl- butanoic acid, (+–)	$C_5H_{10}O_2$	348	[43]
1153	3-Methyl- butanoic acid	$C_5H_{10}O_2$	350	[43]
1154	Acetic Acid, 1-methylethyl ester	$C_5H_{10}O_2$	275	[43]
1155	Acetic acid, 2-propenyl ester	$C_5H_8O_2$	295.372	[43]
1156	Acetic acid, cyano-, ethyl ester	$C_5H_7NO_2$	383.15	[43]
1157	2- Methyl -,(Z)- 2-butenedioic acid	$C_5H_6O_4$	493	[43]
1158	2-Propenoic acid, 2-hydroxylethyl ester	$C_5H_8O_3$	371.15	[43]
1159	Acetic acid, propyl ester	$C_5H_{10}O_2$	288	[43]
1160	Acrylic acid, ethyl ester	$C_5H_8O_2$	282.15	[43]
1161	3-Oxo-butanoic acid, methyl ester	$C_5H_8O_3$	343	[43]
1162	Formic acid, 1,1-dimethylethyl ester	$C_5H_{10}O_2$	264.15	[43]
1163	Formic acid, 1-methyl propyl ester	$C_5H_{10}O_2$	274	[43]
1164	Formic acid, 2-methylpropyl ester	$C_5H_{10}O_2$	277	[43]
1165	Formic acid, butyl ester	$C_5H_{10}O_2$	291	[43]
1166	Pentanedioic acid	$C_5H_8O_4$	482	[43]

(continued)

(continued)

No.	Compound	Formula	Exp.	Ref.
1167	Pentanoic acid	$C_5H_{10}O_2$	358.15	[43]
1168	Pentanoic Acid, 4-oxo	$C_5H_8O_3$	410	[43]
1169	Propanedioic acid, dimethyl ester	$C_5H_8O_4$	363.15	[43]
1170	Propanoic acid, 2,2-dimethyl-	$C_5H_{10}O_2$	336.15	[43]
1171	Propanoic acid, 2-hydroxy-, ethyl ester	$C_5H_{10}O_3$	319.15	[43]
1172	2-Methyl-propanoic acid, methyl ester	$C_5H_{10}O_2$	276.15	[43]
1173	Propanoic acid, ethenyl ester	$C_5H_8O_2$	274	[43]
1174	Propanoic acid, ethyl ester	$C_5H_{10}O_2$	285	[43]
1175	2-Propenoic acid, 2-methyl-, methyl ester	$C_5H_8O_2$	284.15	[43]
1176	Propanoic acid, propyl ester	$C_6H_{12}O_2$	292.15	[43]
1177	Butanoic acid, ethyl ester	$C_6H_{12}O_2$	297.15	[43]
1178	2,2-Dimethyl- butanoic acid	$C_6H_{12}O_2$	352.15	[43]
1179	Butanoic acid, 2-ethyl-	$C_6H_{12}O_2$	360	[43]
1180	Acetic Acid, 1,1-dimethylethyl ester	$C_6H_{12}O_2$	287.25	[43]
1181	Acetic acid, 2-methylpropyl ester	$C_6H_{12}O_2$	291	[43]
1182	Acetic acid, butyl Ester	$C_6H_{12}O_2$	295	[43]
1183	1,1,1,3,3,3-Hexamethyldisilazane	$C_6H_{19}NSi_2$	281.15	[43]
1184	2-Hydroxy-1,2,3-propanetricarboxylic acid	$C_6H_8O_7$	549	[43]
1185	2-Propenoic acid, 1-methylethyl ester	$C_6H_{10}O_2$	281	[43]
1186	2-Propenoic acid, 2 methyl-2-hydroxyethyl ester	$C_6H_{10}O_3$	370.15	[43]
1187	2-Propenoic acid, 2-carboxyethyl ester	$C_6H_8O_4$	419	[43]
1188	2- Butenedioic Acid (Z)-, dimethyl ester	$C_6H_8O_4$	386.15	[43]
1189	2-Propenoic acid, 2-hydroxypropyl ester	$C_6H_{10}O_3$	372.15	[43]
1190	2-Propenoic acid, propyl ester	$C_6H_{10}O_2$	293	[43]
1191	3-Oxo- butanoic acid, ethyl ester	$C_6H_{10}O_3$	330.372	[43]
1192	Formic acid, pentyl ester	$C_6H_{12}O_2$	300	[43]
1193	Ethanedioic acid, diethyl ester	$C_6H_{10}O_4$	339.25	[43]
1194	6-Hydroxy- hexanoic acid	$C_6H_{12}O_3$	444	[43]
1195	Methacrylic acid, ethyl ester	$C_6H_{10}O_2$	293.15	[43]
1196	Phosphoric acid, triethyl ester	$C_6H_{15}O_4P$	372	[43]
1197	Propanoic acid, 2,2'-oxybis-	$C_6H_{10}O_5$	388	[43]
1198	2-Methyl-propanoic acid, ethyl ester	$C_6H_{12}O_2$	287	[43]
1199	Propanoic acid, anhydride	$C_6H_{10}O_3$	336	[43]
1200	Hexanoic acid	$C_6H_{12}O_2$	375	[43]
1201	2-Propenoic acid, 2 methyl-, 2 hydroxypropyl ester	$C_7H_{12}O_3$	369.15	[43]
1202	2,2-Dimethyl- pentanoic acid	$C_7H_{14}O_2$	366	[43]
1203	Butanoic acid, propyl ester	$C_7H_{14}O_2$	305	[43]
1204	Cyclopentaneacetic acid	$C_7H_{12}O_2$	382.15	[43]
1205	Acetic acid, pentyl ester	$C_7H_{14}O_2$	296.15	[43]
1206	2-Propenoic acid, 1-methyl propyl ester	$C_7H_{12}O_2$	299.6	[43]
1207	2-Propenoic acid, 2-methyl-, 2-propenyl ester	$C_7H_{10}O_2$	306.15	[43]
1208	2-Propenoic acid, 2-methyl-, propyl ester	$C_7H_{12}O_2$	304	[43]

(continued)

No.	Compound	Formula	Exp.	Ref.
1209	2-Propenoic acid, 2-methylpropyl ester	$C_7H_{12}O_2$	304.261	[43]
1210	2-Propenoic acid, butyl ester	$C_7H_{12}O_2$	312	[43]
1211	Benzoic acid, 2-chloro-	$C_7H_5ClO_2$	435	[43]
1212	Benzoic acid, 2-hydroxy-	$C_7H_6O_3$	430	[43]
1213	Formic acid, hexyl ester	$C_7H_{14}O_2$	317	[43]
1214	Benzoic acid	$C_7H_6O_2$	394.261	[43]
1215	3-Methyl-butanoic acid, ethyl ester	$C_7H_{14}O_2$	298.15	[43]
1216	Formic acid, cyclohexyl ester	$C_7H_{12}O_2$	324.15	[43]
1217	Heptanoic acid	$C_7H_{14}O_2$	382	[43]
1218	2-Methyl- hexanoic acid	$C_7H_{14}O_2$	375	[43]
1219	Propanedioic acid, diethyl ester	$C_7H_{12}O_4$	366.15	[43]
1220	Propanoic acid, 2,2-dimethyl-, ethenyl ester	$C_7H_{12}O_2$	282.15	[43]
1221	2-Methyl-propanoic acid, propyl ester	$C_7H_{14}O_2$	302	[43]
1222	3-Ethoxy- propanoic acid, ethyl ester	$C_7H_{14}O_3$	331.15	[43]
1223	Propanoic acid, butyl ester	$C_7H_{14}O_2$	305.372	[43]
1224	2-Ethyl- hexanoic acid	$C_8H_{16}O_2$	387.15	[43]
1225	Benzoic acid, 2-hydroxy-, methyl ester	$C_8H_8O_3$	369	[43]
1226	2-Methyl- benzoic acid	$C_8H_8O_2$	407	[43]
1227	4-Formyl- benzoic acid	$C_8H_6O_3$	548	[43]
1228	Benzoic acid, methyl ester	$C_8H_8O_2$	355.15	[43]
1229	Benzoic acid, 2-formyl-	$C_8H_6O_3$	436	[43]
1230	4-Hydroxymethyl-benzoic acid	$C_8H_8O_3$	495	[43]
1231	Butanoic acid, butyl ester	$C_8H_{16}O_2$	326.483	[43]
1232	Butanedioic acid, diethyl ester	$C_8H_{14}O_4$	363	[43]
1233	Acetic acid, cyclohexyl ester	$C_8H_{14}O_2$	330.15	[43]
1234	Acetic acid, hexyl ester	$C_8H_{16}O_2$	310.15	[43]
1235	1,2- Benzenedicarboxylic acid	$C_8H_6O_4$	477	[43]
1236	1,3- Benzenedicarboxylic acid	$C_8H_6O_4$	601	[43]
1237	*trans*-1,4- Cyclohexane Dicarboxylic Acid	$C_8H_{12}O_4$	508.15	[43]
1238	2-Butanedioic acid(Z)-, diethyl ester	$C_8H_{12}O_4$	366.15	[43]
1239	2-Propenoic acid, 1,2-ethanediyl ester	$C_8H_{10}O_4$	373.15	[43]
1240	2-Propenoic acid, 2-methyl-, 2-methyl propyl ester	$C_8H_{14}O_2$	314.15	[43]
1241	2-Propenoic acid, 2-methyl-, butyl ester	$C_8H_{14}O_2$	322	[43]
1242	2-Propenoic acid, 2-methyl-,1,1-dimethylethyl ester	$C_8H_{14}O_2$	300.15	[43]
1243	Butanoic acid, anhydride	$C_8H_{14}O_3$	355	[43]
1244	Formic acid, heptyl ester	$C_8H_{16}O_2$	327	[43]
1245	Formic acid, octyl ester	$C_9H_{18}O_2$	337	[43]
1246	Formic acid, phenylmethyl ester	$C_8H_8O_2$	355	[43]
1247	Formic acid, phenylmethyl ester	$C_8H_8O_2$	355	[43]
1248	Octanedioic acid	$C_8H_{14}O_4$	489	[43]
1249	Octanoic acid	$C_8H_{16}O_2$	393	[43]
1250	2-Methyl-propanoic acid, 2-methylpropyl ester	$C_8H_{16}O_2$	311.15	[43]

(continued)

(continued)

No.	Compound	Formula	Exp.	Ref.
1251	Silicic acid, tetraethyl ester	$C_8H_{20}O_4Si$	324.82	[43]
1252	Octanoic acid, 2-methyl	$C_9H_{18}O_2$	391	[43]
1253	Formic acid, octyl ester	$C_9H_{18}O_2$	337	[43]
1254	Propanoic acid, 2-(2-hydroxy-1-oxopropoxy)-1-carboxyethyl E	$C_9H_{14}O_7$	496	[43]
1255	Benzoic acid, 2-(acetyloxy)	$C_9H_8O_4$	436	[43]
1256	Benzoic acid, 4-formyl-, methyl ester	$C_9H_8O_3$	403	[43]
1257	Benzoic acid, 4-methyl-, methyl ester	$C_9H_{10}O_2$	363.71	[43]
1258	Benzoic acid, ethyl ester	$C_9H_{10}O_2$	357.15	[43]
1259	5-Isobenzofurancarboxylic acid, 1,3-dihydro-1,3-dioxo-	$C_9H_4O_5$	512	[43]
1260	(E)- 3-Phenyl-,2-propenoic acid	$C_9H_8O_2$	445	[43]
1261	Acetic acid, heptyl ester	$C_9H_{18}O_2$	340	[43]
1262	Acetic acid, phenylmethyl ester	$C_9H_{10}O_2$	363.15	[43]
1263	1,2,4- Benzenetricarboxylic acid	$C_9H_6O_6$	550	[43]
1264	Benzeneacetic acid, 4-methoxy-	$C_9H_{10}O_3$	437	[43]
1265	Nonanedioic acid	$C_9H_{16}O_4$	492	[43]
1266	Nonanoic acid	$C_9H_{18}O_2$	401	[43]
1267	Pentanoic acid, butyl ester	$C_9H_{18}O_2$	336	[43]
1268	Acetic acid, 2-ethylhexyl ester	$C_{10}H_{20}O_2$	344.15	[43]
1269	Acetic acid, octyl ester	$C_{10}H_{20}O_2$	359.15	[43]
1270	1,2,4,5-Benzenetetracarboxylic acid	$C_{10}H_6O_8$	597	[43]
1271	1,2- Benzenedicarboxylic acid, dimethyl ester	$C_{10}H_{10}O_4$	419	[43]
1272	1,3- Benzenedicarboxylic acid, dimethyl ester	$C_{10}H_{10}O_4$	411.15	[43]
1273	1,4- Benzenedicarboxylic acid, dimethyl ester	$C_{10}H_{10}O_4$	414.15	[43]
1274	1,4- Cyclohexane Dicarboxylic acid, dimethyl ester	$C_{10}H_{16}O_4$	383.15	[43]
1275	2- Butenedioic acid (Z)-, di-2-propenyl ester	$C_{10}H_{12}O_4$	393	[43]
1276	2- Butenedioic Acid (Z)-, dipropyl ester	$C_{10}H_{16}O_4$	391	[43]
1277	Benzoic acid, propyl ester	$C_{10}H_{12}O_2$	371.15	[43]
1278	3-Methyl-butanoic acid, 3-methylbutyl ester	$C_{10}H_{20}O_2$	345	[43]
1279	Decanedioic acid	$C_{10}H_{18}O_4$	496	[43]
1280	3-Hydroxy-2,2-dimethyl- propanoic acid, 3-hydroxy-2,2-dimethylpropyl ester	$C_{10}H_{20}O_4$	426	[43]
1281	Decanoic acid	$C_{10}H_{20}O_2$	412	[43]
1282	Formic acid, nonyl ester	$C_{10}H_{20}O_2$	353	[43]
1283	Benzoic acid, butyl ester	$C_{11}H_{14}O_2$	379.15	[43]
1284	2-Propenoic acid, 2-ethylhexyl ester	$C_{11}H_{20}O_2$	355.15	[43]
1285	Acetic acid, nonyl ester	$C_{11}H_{22}O_2$	372.15	[43]
1286	Decanoic acid, methyl ester	$C_{11}H_{22}O_2$	367.15	[43]
1287	Formic acid, decyl ester	$C_{11}H_{22}O_2$	367	[43]
1288	Undecanoic acid	$C_{11}H_{22}O_2$	421.15	[43]
1289	Dodecanoic acid	$C_{12}H_{24}O_2$	430	[43]
1290	Acetic acid, decyl ester	$C_{12}H_{24}O_2$	377	[43]

(continued)

No.	Compound	Formula	Exp.	Ref.
1291	Boric acid (H_3BO_3),tributyl ester	$C_{12}H_{27}BO_3$	366	[43]
1292	1,2,4- Benzenetricarboxylic acid, trimethyl ester	$C_{12}H_{12}O_6$	468	[43]
1293	1,2- Benzenedicarboxylic acid, diethyl ester	$C_{12}H_{14}O_4$	390.15	[43]
1294	1,4-Benzenedicarboxylic acid, bis(2-hydroxyethyl) ester	$C_{12}H_{14}O_6$	545	[43]
1295	2,6- Naphthalenedicarboxylic acid	$C_{12}H_8O_4$	544	[43]
1296	2- Butenedioic acid (Z)-, dibutyl ester	$C_{12}H_{20}O_4$	413	[43]
1297	Propanoic acid,2-methyl-,3-hydroxy-2,2, 4-Trimethylpentyl ester	$C_{12}H_{24}O_3$	379.15	[43]
1298	Benzeneacetic acid, alpha-methyl-4- (2-methylpropyl)	$C_{13}H_{18}O_2$	436	[43]
1299	Dodecanoic acid, methyl ester	$C_{13}H_{26}O_2$	391	[43]
1300	Nonanoic acid, butyl ester	$C_{13}H_{26}O_2$	361	[43]
1301	Tridecanoic acid	$C_{13}H_{26}O_2$	438	[43]
1302	Benzoic acid, phenylmethyl ester	$C_{14}H_{12}O_2$	418.15	[43]
1303	1,2- Benzenedicarboxylic acid, dipropyl ester	$C_{14}H_{18}O_4$	433	[43]
1304	1,4- Benzenedicarboxylic acid, bis(2,3-Dihydroxypropyl) ester	$C_{14}H_{18}O_8$	575	[43]
1305	2,6-Naphthalene dicarboxylic acid, dimethyl ester	$C_{14}H_{12}O_4$	479	[43]
1306	Tetradecanoic acid	$C_{14}H_{28}O_2$	446	[43]
1307	Pentadecanoic acid	$C_{15}H_{30}O_2$	454	[43]
1308	1,2- Benzenedicarboxylic acid, bis(2-methylpropyl) ester	$C_{16}H_{22}O_4$	434.15	[43]
1309	1,2- Benzenedicarboxylic dibutyl ester acid	$C_{16}H_{22}O_4$	430.15	[43]
1310	Hexadecanoic acid	$C_{16}H_{32}O_2$	461	[43]
1311	2-Methyl- propanoic acid, 2,2-dimethyl-1- (1-methyl ethyl)-1,3-propanediyl ester	$C_{16}H_{30}O_4$	394	[43]
1312	Tetradecanoic acid, 1-methylethyl ester	$C_{17}H_{34}O_2$	423	[43]
1313	Heptadecanoic acid	$C_{17}H_{34}O_2$	469	[43]
1314	(Z,Z,Z)- 9,12,15-Octadecatrienoic acid	$C_{18}H_{30}O_2$	471	[43]
1315	(Z,Z)- 9,12-Octadecadienoic acid	$C_{18}H_{32}O_2$	467	[43]
1316	(Z)- 9-Octadecenoic acid	$C_{18}H_{34}O_2$	462	[43]
1317	Decanedioic acid, dibutyl ester	$C_{18}H_{34}O_4$	451.15	[43]
1318	Hexanedioic acid, dihexyl ester	$C_{18}H_{34}O_4$	464	[43]
1319	Phosphoric acid, triphenyl ester	$C_{18}H_{15}O_4P$	493.15	[43]
1320	Octadecanoic acid	$C_{18}H_{36}O_2$	476.15	[43]
1321	(Z)- 9-Octadecenoic acid, methyl ester	$C_{19}H_{36}O_2$	450	[43]
1322	Nonadecanoic acid	$C_{19}H_{38}O_2$	483	[43]
1323	1-Phenanthrenecarboxylic acid, 1,2,3,4,4a,9,10,10a-octahydro-1,4a-dimethyl-7-(1- methylethyl)-,[1R-(1alpha, 4abeta, 10aalpha)]-	$C_{20}H_{28}O_2$	470	[43]

(continued)

(continued)

No.	Compound	Formula	Exp.	Ref.
1324	1,2,3,4,4a,4b,5,6,7,8,10,10a-Dodecahydro-1,4a,7-Trimethyl-1-Phenanthrenecarboxylic acid	$C_{20}H_{30}O_2$	447	[43]
1325	1,2- Benzenedicarboxylic acid, dihexyl ester	$C_{20}H_{30}O_4$	466.15	[43]
1326	7-Ethenyl-1,2,3,4,4a,4b,5,6,7,9,10-10a-dode-cahydro-1,4a,7-trimethyl-1-phenanthrenecar-boxylic acid	$C_{20}H_{30}O_2$	447	[43]
1327	2-Propenoic acid, 2-methyl-,hexadecyl ester	$C_{20}H_{38}O_2$	452	[43]
1328	Eicosanic acid	$C_{20}H_{40}O_2$	490	[43]
1329	Phosphoric acid, tris(2-methylphenyl) ester	$C_{21}H_{21}O_4P$	498.15	[43]
1330	1,2- Benzenedicarboxylic acid, diheptyl ester	$C_{22}H_{34}O_4$	485	[43]
1331	Hexanedioic acid, bis(2-ethylhexyl)ester	$C_{22}H_{42}O_4$	466	[43]
1332	Hexanedioic acid, dioctyl ester	$C_{22}H_{42}O_4$	500.15	[43]
1333	Octadecanoic acid, butyl ester	$C_{22}H_{44}O_2$	433.15	[43]
1334	1,2- Benzenedicarboxylic acid, bis(2-ethyl-hexyl) ester	$C_{24}H_{38}O_4$	489	[43]
1335	1,2- Benzenedicarboxylic acid, diisooctyl ester	$C_{24}H_{38}O_4$	504	[43]
1336	1,2- Benzenedicarboxylic acid, heptyl, nonyl ester	$C_{24}H_{38}O_4$	500	[43]
1337	1,4- Benzenedicarboxylic acid, dioctyl ester	$C_{24}H_{38}O_4$	507	[43]
1338	1,2-Benzene dicarboxylic acid, heptyl, undecyl ester	$C_{26}H_{42}O_4$	514	[43]
1339	Diisononyl ester -1,2- benzenedicarboxylic acid	$C_{26}H_{42}O_4$	489	[43]
1340	1,2- Benzenedicarboxylic acid, dinonyl ester	$C_{26}H_{42}O_4$	489.15	[43]
1341	1,2- Benzenedicarboxylic acid, didecyl ester	$C_{28}H_{46}O_4$	505	[43]
1342	1,2- Benzenedicarboxylic acid, diisodecyl ester	$C_{28}H_{46}O_4$	505.372	[43]
1343	1,2- Benzenedicarboxylic acid, nonyl undecyl ester	$C_{28}H_{46}O_4$	524	[43]
1344	1,2,4- Benzenetricarboxylic acid, triheptyl ester	$C_{30}H_{48}O_6$	546	[43]
1345	1,2- Benzenedicarboxylic acid, diundecyl ester	$C_{30}H_{50}O_4$	535	[43]
1346	Methyltrichlorosilane	CH_3Cl_3Si	276.15	[258]
1347	Trichloro(Chloromethyl) silane	CH_3Cl_4Si	342.15	[258]
1348	Methyldichlorosilane	CH_4Cl_2Si	241.15	[258]
1349	Vinyltrichlorosilane	$C_2H_3Cl_3Si$	294.15	[258]
1350	Dichloromethylmethyldichlorosilane	$C_2H_4Cl_4Si$	301.15	[258]
1351	Dichloroethylsilane	$C_2H_6Cl_2Si$	263.75	[258]
1352	Dimethyldichlorosilane	$C_2H_6Cl_2Si$	257.15	[258]
1353	Ethyltrichlorosilane	$C_2H_5Cl_3Si$	263.15	[258]
1354	Dimethylchlorosilane	C_2H_7ClSi	245.15	[258]
1355	Ethylsilane	C_2H_8Si	213.15	[258]
1356	1,1,1-Trifluoropropyltrichlorosilane	$C_3H_5Cl_3F_3Si$	288.15	[258]
1357	Allyltrichlorosilane	$C_3H_5Cl_3Si$	304.15	[258]

(continued)

No.	Compound	Formula	Exp.	Ref.
1358	Methylvinyldichlorosilane	$C_3H_6Cl_2Si$	272.15	[258]
1359	3-Chloropropyltrichlorosilane	$C_3H_6Cl_4Si$	357.15	[258]
1360	n-Propyltrichlorosilane	$C_3H_7Cl_3Si$	310.15	[258]
1361	Bromomethyldimethylchlorosilane	$C_3H_8BrClSi$	314.15	[258]
1362	Chloromethyldimethylchlorosilane	$C_3H_8Cl_2Si$	295.15	[258]
1363	Iodotrimethylsilane	C_3H_9ISi	288.15	[258]
1364	Trimethylbromosilane	C_3H_9BrSi	270.15	[258]
1365	Trimethylchlorosilane	C_3H_9ClSi	255.15	[258]
1366	Trimethylfluorosilane	C_3H_9FSi	238.15	[258]
1367	Trimethylsilane	$C_3H_{10}Si$	204.15	[258]
1368	Trimethoxysilane	$C_3H_{10}O_3Si$	264.15	[258]
1369	Trichloro(3-cyanopropyl) silane	$C_4H_6Cl_3NSi$	365.15	[258]
1370	Cyanoethylmethyldichlorosilane	$C_4H_7Cl_2NSi$	333.15	[258]
1371	Dichloromethyl-3,3,3-trifluoropropylsilane	$C_4H_7Cl_2F_3Si$	288.15	[258]
1372	Allylmethyldichlorosilane	$C_4H_8Cl_2Si$	300.15	[258]
1373	Dimethylvinylchlorosilane	C_4H_9ClSi	268.15	[258]
1374	3-Chloropropylmethyldichlorosilane	$C_4H_9Cl_3Si$	359.15	[258]
1375	t-Butyltrichlorosilane	$C_4H_9Cl_3Si$	294.15	[258]
1376	Isobutyltrichlorosilane	$C_4H_9Cl_3Si$	310.15	[258]
1377	n-Butyltrichlorosilane	$C_4H_9Cl_3Si$	327.15	[258]
1378	Trimethylsilylisothiocyanate	C_4H_9NSSi	308.15	[258]
1379	n-Propylmethyldichlorosilane	$C_4H_{10}Cl_2Si$	300.15	[258]
1380	Bromomethyltrimethylsilane	$C_4H_{11}BrSi$	289.15	[258]
1381	Ethyldimethylchlorosilane	$C_4H_{11}ClSi$	269.15	[258]
1382	1,3-Dichlorotetramethyldisiloxane	$C_4H_{12}Cl_2OSi_2$	288.15	[258]
1383	Methoxytrimethylsilane	$C_4H_{12}OSi$	243.15	[258]
1384	Dimethyldimethoxysilane	$C_4H_{12}O_2Si$	265.15	[258]
1385	Methyltrimethoxysilane	$C_4H_{12}O_3Si$	284.15	[258]
1386	Tetramethoxysilane	$C_4H_{12}O_4Si$	294.15	[258]
1387	Tetramethylsilane	$C_4H_{12}Si$	245.15	[258]
1388	Diethylsilane	$C_4H_{12}Si$	253.15	[258]
1389	(N,N-Dimethylamino)dimethylsilane	$C_4H_{13}NSi$	260.15	[258]
1390	1,1,3,3-Tetramethyldisiloxane	$C_4H_{14}OSi_2$	251.15	[258]
1391	Tetramethyldisilazane	$C_4H_{15}NSi_2$	284.15	[258]
1392	1,3,5,7-Tetramethylcyclotetrasiloxane	$C_4H_{16}O_4Si_4$	304.15	[258]
1393	Tetraisocyanatosilane	$C_4N_4O_4Si$	323.15	[258]
1394	Cyclopentyltrichlorosilane	$C_5H_9Cl_3Si$	350.15	[258]
1395	Trimethylsilyltrifluoroacetate	$C_5H_9F_3O_2Si$	273.15	[258]
1396	Chlorodimethyl(3,3,3-trifluoropropyl) silane	$C_5H_{10}ClF_3Si$	293.15	[258]
1397	Cyclopentyldichlorosilane	$C_5H_{10}Cl_2Si$	316.15	[258]
1398	n-Amyltrichlorosilane	$C_5H_{11}Cl_3Si$	303.15	[258]
1399	3-Chloro-2-methylpropylmethyldichlorosilane	$C_5H_{11}Cl_3Si$	363.15	[258]
1400	Allyldimethylchlorosilane	$C_5H_{11}ClSi$	278.15	[258]
1401	sec-Amyltrichlorosilane	$C_5H_{11}Cl_3Si$	303.15	[258]

(continued)

(continued)

No.	Compound	Formula	Exp.	Ref.
1402	n-Butylmethyldichlorosilane	$C_5H_{12}Cl_2Si$	303.15	[258]
1403	Trimethyl(vinyloxy)silane	$C_5H_{12}OSi$	263.15	[258]
1404	Methylvinyldimethoxysilane	$C_5H_{12}O_2Si$	277.15	[258]
1405	Vinyltrimethoxysilane	$C_5H_{12}O_3Si$	296.15	[258]
1406	Trimethylvinylsilane	$C_5H_{12}Si$	239.15	[258]
1407	Cyclotrimethylenedimethylsilane	$C_5H_{12}Si$	268.15	[258]
1408	Diethylmethylchlorosilane	$C_5H_{13}ClSi$	288.15	[258]
1409	Chloromethyldimethylethoxysilane	$C_5H_{13}ClOSi$	299.15	[258]
1410	Ethoxytrimethylsilane	$C_5H_{14}OSi$	246.15	[258]
1411	Ethyltrimethoxysilane	$C_5H_{14}O_3Si$	300.15	[258]
1412	(N,N-Dimethylamino)trimethylsilane	$C_5H_{15}NSi$	254.15	[258]
1413	Bis(dimethylamino)methylsilane	$C_5H_{16}N_2Si$	270.15	[258]
1414	Pentamethyldisiloxane	$C_5H_{16}OSi_2$	254.15	[258]
1415	Trichloro(dichlorophenyl)silane	$C_6H_3Cl_5Si$	419.15	[258]
1416	Trichloro(p-chlorophenyl)silane	$C_6H_4Cl_4Si$	397.05	[258]
1417	Phenyltrichlorosilane	$C_6H_5Cl_3Si$	364.15	[258]
1418	Phenyltriflurosilane	$C_6H_5F_3Si$	268.15	[258]
1419	Phenyldichlorosilane	$C_6H_6Cl_2Si$	333.15	[258]
1420	Phenylsilane	C_6H_8Si	281.15	[258]
1421	1-Hexenyltrichlorosilane	$C_6H_{11}Cl_3Si$	333.15	[258]
1422	Cyclohexyltrichlorosilane	$C_6H_{11}Cl_3Si$	358.15	[258]
1423	Dimethyldiacetoxysilane	$C_6H_{12}O_4Si$	310.15	[258]
1424	1-Trimethylsilylpropyne	$C_6H_{12}Si$	270.15	[258]
1425	N-Methyl-N-trimethylsilyltrifluroacetamide	$C_6H_{12}F_3NOSi$	298.15	[258]
1426	Propargyloxytrimethylsilane	$C_6H_{12}OSi$	300.15	[258]
1427	Divinyldimethylsilane	$C_6H_{12}Si$	281.15	[258]
1428	n-Hexyltrichlorosilane	$C_6H_{13}Cl_3Si$	358.15	[258]
1429	Hexyltriflurosilane	$C_6H_{13}F_3Si$	266.15	[258]
1430	Vinyldimethylethoxysilane	$C_6H_{14}OSi$	277.15	[258]
1431	Trimethyl[(1-methylethenyl)oxy]silane	$C_6H_{14}OSi$	272.15	[258]
1432	Allyloxytrimethylsilane	$C_6H_{14}OSi$	273.15	[258]
1433	Allyltrimethoxysilane	$C_6H_{14}O_3Si$	319.15	[258]
1434	n-Amylmethyldichlorosilane	$C_6H_{14}Cl_2Si$	308.15	[258]
1435	Hexyldichlorosilane	$C_6H_{14}Cl_2Si$	337.15	[258]
1436	1,1-Dimethyl-1-sila-2-oxacyclohexane	$C_6H_{14}OSi$	287.15	[258]
1437	Acetoxymethyltrimethylsilane	$C_6H_{14}O_2Si$	293.15	[258]
1438	Allyltrimethylsilane	$C_6H_{14}Si$	280.15	[258]
1439	Cyclotetramethylenedimethylsilane	$C_6H_{14}Si$	288.15	[258]
1440	[2-(chloromethoxy)ethyl]trimethylsilane	$C_6H_{15}ClOSi$	319.15	[258]
1441	3-Chloropropyltrimethoxysilane	$C_6H_{15}ClO_3Si$	348.15	[258]
1442	Triethylchlorosilane	$C_6H_{15}ClSi$	303.15	[258]
1443	t-Butyldimethylchlorosilane	$C_6H_{15}ClSi$	295.15	[258]
1444	3-Chloropropylmethyldimethoxysilane	$C_6H_{15}ClO_2Si$	318.15	[258]
1445	n-Propyltrimethoxysilane	$C_6H_{16}O_3Si$	298.15	[258]

(continued)

No.	Compound	Formula	Exp.	Ref.
1446	Bis(chloromethyl)tetramethyldisiloxane	$C_6H_{16}Cl_2OSi_2$	346.15	[258]
1447	Trimethyl-n-propoxysilane	$C_6H_{16}OSi$	283.15	[258]
1448	t-Butyldimethylsilanol	$C_6H_{16}OSi$	318.15	[258]
1449	Dimethyldiethoxysilane	$C_6H_{16}O_2Si$	284.15	[258]
1450	Triethylsilane	$C_6H_{16}Si$	270.15	[258]
1451	Hexylsilane	$C_6H_{16}Si$	285.15	[258]
1452	t-Butyldimethylsilane	$C_6H_{16}Si$	251.15	[258]
1453	Hexamethyldisiloxane	$C_6H_{18}OSi_2$	270.15	[258]
1454	Hexamethyldisilane	$C_6H_{18}Si_2$	272.15	[258]
1455	Bis(dimethylamino)dimethylsilane	$C_6H_{18}N_2Si$	266.15	[258]
1456	Hexamethylcyclotrisiloxane	$C_6H_{18}O_3Si_3$	305.15	[258]
1457	Hexamethyldisilthiane	$C_6H_{18}SSi_2$	299.15	[258]
1458	Hexamethyldisilazane	$C_6H_{19}NSi_2$	281.15	[258]
1459	Tris(dimethylamino)silane	$C_6H_{19}N_3Si$	298.15	[258]
1460	Hexamethyltrisiloxane	$C_6H_{20}O_2Si_3$	293.15	[258]
1461	p-Tolytrichlorosilane	$C_7H_7Cl_3Si$	365.15	[258]
1462	Phenylmethyldichlorosilane	$C_7H_8Cl_2Si$	340.15	[258]
1463	Phenylmethylchlorosilane	C_7H_9ClSi	355.15	[258]
1464	Methyltriacetoxysilane	$C_7H_{12}O_6Si$	358.15	[258]
1465	Trivinylmethylsilane	$C_7H_{12}Si$	288.15	[258]
1466	Cyclohexylmethyldichlorosilane	$C_7H_{14}Cl_2Si$	355.15	[258]
1467	1-(Trimethylsiloxy)-1,3-butadiene	$C_7H_{14}OSi$	298.15	[258]
1468	2-Chloromethyl-3-trimethylsilyl-1-propene	$C_7H_{15}ClSi$	316.15	[258]
1469	n-Heptyltrichlorosilane	$C_7H_{15}Cl_3Si$	337.15	[258]
1470	Ethyl(2-trimethylsilyl)acetate	$C_7H_{15}O_2Si$	308.15	[258]
1471	Vinyldiethylmethylsilane	$C_7H_{16}Si$	288.15	[258]
1472	n-Hexylmethyldichlorosilane	$C_7H_{16}Cl_2Si$	358.15	[258]
1473	Vinylmethyldiethoxysilane	$C_7H_{16}O_2Si$	289.15	[258]
1474	Cyclopentamethylenedimethylsilane	$C_7H_{16}Si$	303.15	[258]
1475	N-(Trimethylsilyl)morpholine	$C_7H_{17}NOSi$	300.15	[258]
1476	Isobutyltrimethoxysilane	$C_7H_{18}O_3Si$	315.15	[258]
1477	Bis(trimethylsily)carbodiimide	$C_7H_{18}N_2Si_2$	315.15	[258]
1478	t-Butoxytrimethylsilane	$C_7H_{18}OSi$	287.15	[258]
1479	Vinylpentamethyldisiloxane	$C_7H_{18}OSi_2$	285.15	[258]
1480	(1-Methoxy-2-propoxy)trimethylsilane	$C_7H_{18}O_2Si$	293.15	[258]
1481	Methyltriethoxysilane	$C_7H_{18}O_3Si$	297.15	[258]
1482	(Diethylamino)trimethylsilane	$C_7H_{19}NSi$	283.15	[258]
1483	Bis(trimethylsilyl)methane	$C_7H_{20}Si_2$	285.15	[258]
1484	Heptamethyldisilazane	$C_7H_{21}NSi_2$	300.15	[258]
1485	Heptamethylsiloxane	$C_7H_{21}OSi_2$	296.15	[258]
1486	Trichloro(2-phenylethyl)silane	$C_8H_9Cl_3Si$	364.15	[258]
1487	Phenylethyldichlorosilane	$C_8H_{10}Cl_2Si$	365.15	[258]
1488	Dimethyl(pentafluorophenyl) silane	$C_8H_{10}F_5Si$	313.15	[258]
1489	Phenyldimethylchlorosilane	$C_8H_{11}ClSi$	334.15	[258]

(continued)

(continued)

No.	Compound	Formula	Exp.	Ref.
1490	Vinyltriacetoxysilane	$C_8H_{12}O_6Si$	361.15	[258]
1491	Tetravinylsilane	$C_8H_{12}Si$	291.15	[258]
1492	Phenyldimethylsilane	$C_8H_{12}Si$	311.15	[258]
1493	Cyclopentadienyltrimethylsilane	$C_8H_{14}Si$	302.15	[258]
1494	Ethyltriacetoxysilane	$C_8H_{14}O_6Si$	377.15	[258]
1495	3-(Trimethylsilyl)cyclopentene	$C_8H_{16}Si$	299.15	[258]
1496	(Cyclopentenyloxy)trimethylsilane	$C_8H_{16}OSi$	311.15	[258]
1497	(2,4-Pentadienyl)trimethylsilane	$C_8H_{16}Si$	291.15	[258]
1498	n-Octyltrichlorosilane	$C_8H_{17}Cl_3Si$	369.15	[258]
1499	Vinyltriethylsilane	$C_8H_{18}Si$	318.15	[258]
1500	n-Heptylmethyldichlorosilane	$C_8H_{18}Cl_2Si$	339.15	[258]
1501	Tetramethyldivinyldisiloxane	$C_8H_{18}OSi_2$	292.15	[258]
1502	Triethylacetoxysilane	$C_8H_{18}O_2Si$	314.15	[258]
1503	Vinyltriethoxysilane	$C_8H_{18}O_3Si$	307.15	[258]
1504	1,3-Divinyltetramethyldisilazane	$C_8H_{19}NSi_2$	307.15	[258]
1505	n-Octylsilane	$C_8H_{20}Si$	309.15	[258]
1506	Diethyldiethoxysilane	$C_8H_{20}O_2Si$	316.15	[258]
1507	Tetraethoxysilane	$C_8H_{20}O_4Si$	323.15	[258]
1508	Tetraethylsilane	$C_8H_{20}Si$	305.15	[258]
1509	(N,N-Dimethylamino)triethylsilane	$C_8H_{21}NSi$	299.15	[258]
1510	Aminoethylaminopropyltrimethoxysilane	$C_8H_{22}N_2O_3Si$	369.15	[258]
1511	Tetramethyldiethoxysiloxane	$C_8H_{22}O_3Si_2$	311.15	[258]
1512	Octamethyltrisiloxane	$C_8H_{24}O_2Si_3$	303.15	[258]
1513	Octamethylcyclotetrasiloxane	$C_8H_{24}O_4Si_4$	330.15	[258]
1514	Octamethylcyclotetrasilazane	$C_8H_{28}N_4Si_4$	339.15	[258]
1515	Tetrakisdimethylsiloxysilane	$C_8H_{28}O_4Si_5$	340.15	[258]
1516	p-Tolydimethylchlorosilane	$C_9H_{13}ClSi$	340.15	[258]
1517	p-Chlorophenyltrimethylsilane	$C_9H_{13}ClSi$	322.15	[258]
1518	Methylphenyldimethoxysilane	$C_9H_{14}O_2Si$	349.15	[258]
1519	Phenyltrimethoxysilane	$C_9H_{14}O_3Si$	358.15	[258]
1520	Trimethylphenylsilane	$C_9H_{14}Si$	313.15	[258]
1521	n-(t-Butyldimethylsilyl)-n-methyl-trifluoroacetamide	$C_9H_{18}F_3NOSi$	325.15	[258]
1522	2-Cyanoethyltriethoxysilane	$C_9H_{19}NO_3Si$	359.15	[258]
1523	Trichlorononylsilane	$C_9H_{19}Cl_3Si$	387.15	[258]
1524	n-Otadecyltrichlorosilane	$C_9H_{19}Cl_3OSi$	466.15	[258]
1525	3-Glycidoxypropyltrimethoxysilane	$C_9H_{20}O_5Si$	383.15	[258]
1526	Cyclohexylmethyldimethoxysilane	$C_9H_{20}O_2Si$	343.15	[258]
1527	Di-t-butylmethylchlorosilane	$C_9H_{21}ClSi$	342.15	[258]
1528	Tri-n-propylchlorosilane	$C_9H_{21}ClSi$	349.15	[258]
1529	Hexyltrimethoxysilane	$C_9H_{22}O_3Si$	335.15	[258]
1530	Triisopropylsilane	$C_9H_{22}Si$	310.15	[258]
1531	Tri-n-propylsilane	$C_9H_{22}Si$	316.15	[258]
1532	(Diisopropylamino)trimethylsilane	$C_9H_{23}NSi$	301.15	[258]

(continued)

No.	Compound	Formula	Exp.	Ref.
1533	Vinylmethylbis (trimethylsiloxy) silane	$C_9H_{24}O_2Si_3$	324.15	[258]
1534	1-Chlorononamethyltetrasiloxane	$C_9H_{27}ClO_3Si_4$	347.15	[258]
1535	Tris(trimethylsiloxy) boron	$C_9H_{27}BO_3Si_3$	315.15	[258]
1536	Propyltris(dimethylsiloxy)silane	$C_9H_{28}O_3Si_4$	323.15	[258]
1537	Tris(trimethylsiloxy)silane	$C_9H_{28}O_3Si_4$	323.15	[258]
1538	Tris(trimethylsilyl)silane	$C_9H_{28}Si_4$	328.15	[258]
1539	Benzyltrimethylsilane	$C_{10}H_{16}Si$	335.15	[258]
1540	3-Methacryloxypropyltrimethoxysilane	$C_{10}H_{20}O_5Si$	365.15	[258]
1541	1-Heptynyltrimethylsilane	$C_{10}H_{20}Si$	315.15	[258]
1542	1,4-Divinyltetramethyldisilylethane	$C_{10}H_{22}Si_2$	313.15	[258]
1543	n-Octyldimethylchlorosilane	$C_{10}H_{23}ClSi$	370.15	[258]
1544	Bis(diethylamino)dimethylsilane	$C_{10}H_{26}N_2Si$	308.15	[258]
1545	Tristrimethylsiloxymethylsilane	$C_{10}H_{30}O_3Si_4$	336.15	[258]
1546	Decamethyltetrasiloxane	$C_{10}H_{30}O_3Si_4$	335.15	[258]
1547	Decamethylcyclopentasiloxane	$C_{10}H_{30}O_5Si_5$	350.15	[258]
1548	2-(3,4-epoxycyclohexyl)ethyltrimethoxysilane	$C_{11}H_{22}O_4Si$	419.15	[258]
1549	n-Decylmethyldichlorosilane	$C_{11}H_{24}Cl_2Si$	393.15	[258]
1550	Vinyltris(methoxyethoxy)silane	$C_{11}H_{24}O_6Si$	353.15	[70]
1551	Dichlorosilane	$C_{12}H_2Si$	236.15	[258]
1552	Diphenyldichlorosilane	$C_{12}H_{10}Cl_2Si$	415.15	[258]
1553	Phenyltriethoxysilane	$C_{12}H_{20}O_3Si$	393.15	[258]
1554	Diphenylsilane	$C_{12}H_{12}Si$	371.15	[258]
1555	Tetramethyltetravinylcyclotetrasilane	$C_{12}H_{24}Si_4$	385.15	[258]
1556	n-Dodecyltrichlorosilane	$C_{12}H_{25}Cl_3Si$	435.15	[258]
1557	Tetra-n-propoxysilane	$C_{12}H_{28}O_4Si$	368.15	[258]
1558	Hexaethyldisiloxane	$C_{12}H_{30}OSi_2$	349.15	[258]
1559	Dodecamethylpentasiloxane	$C_{12}H_{36}O_4Si_5$	352.15	[258]
1560	Dodecamethylcyclohexasiloxane	$C_{12}H_{36}O_6Si_6$	366.15	[258]
1561	Methyldiphenylchlorosilane	$C_{13}H_{13}ClSi$	411.15	[258]
1562	Diphenylmethylsilane	$C_{13}H_{14}Si$	393.15	[258]
1563	Dichloromethyl(2-tricyclo[3.3.1.1(3,7)-] dec-1-ylethyl) silane	$C_{13}H_{22}Cl_2Si$	408.15	[258]
1564	n-Dodecylmethyldichlorosilane	$C_{13}H_{28}Cl_2Si$	416.15	[258]
1565	Diphenyldimethoxysilane	$C_{14}H_{16}O_2Si$	394.15	[258]
1566	Octyltriethoxysilane	$C_{14}H_{32}O_3Si$	373.15	[258]
1567	Tetradecamethylhexasiloxane	$C_{14}H_{42}O_5Si_6$	375.15	[258]
1568	Flusilazol	$C_{16}H_{15}F_2N_3Si$	464.15	[258]
1569	Hexadecyltrichlorosilane	$C_{16}H_{33}Cl_3Si$	427.15	[258]
1570	Hexadecamethylheptasiloxane	$C_{16}H_{48}O_6Si_7$	406.15	[258]
1571	Triphenylsilane	$C_{18}H_{16}Si$	416.35	[258]
1572	Octadecamethyloctasiloxane	$C_{18}H_{54}O_7Si_8$	417.15	[258]
1573	Trisilane	H_8Si_3	273.15	[258]
1574	Trichlorosilane	Cl_3HSi	245.15	[258]
1575	Carbon disulfide	CS_2	243.15	[43]

(continued)

(continued)

No.	Compound	Formula	Exp.	Ref.
1576	Carbon oxide sulfide	COS	186	[43]
1577	Methanesulfonyl chloride	CH_3ClO_2S	383.15	[43]
1578	Methanethiol	CH_4S	217	[43]
1579	Thiirane	C_2H_4S	283.15	[43]
1580	Acetic acid, mercapto-	$C_2H_4O_2S$	398	[43]
1581	Ethanesulfonyl chloride	$C_2H_5ClO_2S$	356.48	[43]
1582	Ethanethiol	C_2H_6S	225	[43]
1583	Thiobis-methane	C_2H_6S	237.15	[43]
1584	Dimethyl-disulfide	$C_2H_6S_2$	280	[43]
1585	1,2-Dithio-ethane	$C_2H_6S_2$	317.15	[43]
1586	2-Mercapto-ethanol	C_2H_6OS	340.15	[43]
1587	Sulfinylbis-methane	C_2H_6OS	361	[43]
1588	Sulfuric acid, dimethyl ester	$C_2H_6O_4S$	356.15	[43]
1589	Sulfuric acid, monoethyl ester	$C_2H_6O_4S$	457	[43]
1590	Thietane	C_3H_6S	271.15	[43]
1591	1-Propanesulfonyl chloride	$C_3H_7ClO_2S$	353.15	[43]
1592	1-Propanethiol	C_3H_8S	253.15	[43]
1593	Disulfide, ethyl methyl	$C_3H_8S_2$	299	[43]
1594	(Methylthio)-ethane	C_3H_8S	258.15	[43]
1595	Propane-2-thiol	C_3H_8S	238.15	[43]
1596	Thiophene	C_4H_4S	266.45	[43]
1597	Ethanethioic acid, S-ethyl ester	C_4H_8OS	290	[43]
1598	3-(Methylthio)-propanal	C_4H_8OS	334.15	[43]
1599	Tetrahydro- thiophene	C_4H_8S	285.15	[43]
1600	Tetrahydro-,thiophene, 1,1-dioxide	$C_4H_8O_2S$	446.15	[43]
1601	1-Butanethiol	$C_4H_{10}S$	274.816	[43]
1602	1- Propanethiol, 2- methyl	$C_4H_{10}S$	264.15	[43]
1603	2-Butanethiol	$C_4H_{10}S$	250.15	[43]
1604	Disulfide, diethyl	$C_4H_{10}S_2$	313.15	[43]
1605	2,2'-Oxybis-ethanethiol	$C_4H_{10}OS_2$	371.15	[43]
1606	2-(Ethylthio)-ethanol	$C_4H_{10}OS$	350.15	[43]
1607	Ethanol,2,2'-thiobis-	$C_4H_{10}O_2S$	433.15	[43]
1608	1-(Methylthio)-propane	$C_4H_{10}S$	270	[43]
1609	2-(Methylthio)-propane	$C_4H_{10}S$	262	[43]
1610	1,1'-Thiobis-ethane	$C_4H_{10}S$	263.15	[43]
1611	Sulfuric acid, diethyl ester	$C_4H_{10}O_4S$	351.15	[43]
1612	Sulfurous acid, diethyl ester	$C_4H_{10}O_3S$	326.15	[43]
1613	Tetrachloro-thiophene	C_4Cl_4S	395	[43]
1614	2-Methyl-thiophene	C_5H_6S	280.15	[43]
1615	3-Methyl-thiophene	C_5H_6S	284.15	[43]
1616	Tetrahydro-2-methyl- thiophene,	$C_5H_{10}S$	295	[43]
1617	Tetrahydro-3-methyl- thiophene, 1,1,dioxide	$C_5H_{10}O_2S$	413	[43]
1618	1-Pentanethiol	$C_5H_{12}S$	291	[43]
1619	2-Pentanethiol	$C_5H_{12}S$	282.46	[43]

(continued)

No.	Compound	Formula	Exp.	Ref.
1620	1-(Methylthio)-butane	$C_5H_{12}S$	289	[43]
1621	Disulfide, ethyl propyl	$C_5H_{12}S_2$	324	[43]
1622	2-Methyl-2-(methylthio)-propane	$C_5H_{12}S$	270.15	[43]
1623	Benzenethiol	C_6H_6S	346.15	[43]
1624	2,5-Dimethyl thiophene	C_6H_8S	296.15	[43]
1625	2-Ethyl-thiophene	C_6H_8S	297	[43]
1626	1- Propene 3,3,thiobis	$C_6H_{10}S$	319.15	[43]
1627	1-Hexanethiol	$C_6H_{14}S$	293.15	[43]
1628	2-Methyl-2-(methylthio)-butane	$C_6H_{14}S$	302	[43]
1629	2-(Ethylthio)-2-methyl-propane	$C_6H_{14}S$	282	[43]
1630	1,1′-Thiobis-propane	$C_6H_{14}S$	301	[43]
1631	Disulfide, dipropyl	$C_6H_{14}S_2$	339.15	[43]
1632	1,1′-Sulfonylbis-propane	$C_6H_{14}O_2S$	399.15	[43]
1633	1,2-Benzisothiazol -3(2H)-one-1,1-dioxide	$C_7H_5NO_3S$	513	[43]
1634	Benzene methanethiol	C_7H_8S	343.15	[43]
1635	(Methylthio)-benzene	C_7H_8S	342	[43]
1636	2-Propyl-thiophene	$C_7H_{10}S$	317.59	[43]
1637	1-Heptanethiol	$C_7H_{16}S$	319.15	[43]
1638	Benzo(beta)thiophene	C_8H_6S	383.15	[43]
1639	2,3-Dihydro- benzo [B] thiophene	C_8H_8S	365	[43]
1640	3-Ethyl-2,5-dimethyl-thiophene	$C_8H_{12}S$	330	[43]
1641	1-Octanethiol	$C_8H_{18}S$	341.15	[43]
1642	Propane,2,2 ′-thiobis-2-methyl	$C_8H_{18}S$	300	[43]
1643	1-1′-Thiobis-butane	$C_8H_{18}S$	331	[43]
1644	2,4,4-Trimethyl-2- pentanethiol	$C_8H_{18}S$	304	[43]
1645	Disulfide, bis(1,1-dimethylethyl)	$C_8H_{18}S_2$	337.15	[43]
1646	1,1′-Sulfonylbis-butane	$C_8H_{18}O_2S$	416.15	[43]
1647	2-Methyl-benzo[B]thiophene	C_9H_8S	386.15	[43]
1648	2,3-Dihydro-2- methyl -benzo (B)thiophene	$C_9H_{10}S$	372	[43]
1649	1-Nonanethiol	$C_9H_{20}S$	351.15	[43]
1650	Tert-nonanethiol	$C_9H_{20}S$	338.15	[43]
1651	Benzo[B]thiophene, 2,7-dimethyl	$C_{10}H_{10}S$	382	[43]
1652	1,1′-Thiobis-3-Methyl-butane	$C_{10}H_{22}S$	352	[43]
1653	Sulfide, ethyl octyl	$C_{10}H_{22}S$	366	[43]
1654	1-Decanethiol	$C_{10}H_{22}S$	371.15	[43]
1655	1-Undecanethiol	$C_{11}H_{24}S$	381	[43]
1656	Dibenzothiophene	$C_{12}H_8S$	431	[43]
1657	Disulfide, diphenyl	$C_{12}H_{10}S_2$	436	[43]
1658	1,1′-Thiobis-cyclohexane	$C_{12}H_{22}S$	393	[43]
1659	Tert-dodecanethiol	$C_{12}H_{26}S$	363.15	[43]
1660	4-Methyl-dibenzothiophene	$C_{13}H_{10}S$	447	[43]
1661	4,6-Dimethyl-dibenzothiophene	$C_{14}H_{12}S$	473	[43]
1662	1,1′-Thiobis-octane	$C_{16}H_{34}S$	428	[43]
1663	Hydrogen sulfide	H_2S	167	[43]

Appendix B

The reported autoignition temperature of combustible organic compounds

No.	Compound	Formula	Exp.	Ref.
1	Methane	CH_4	868	[70]
2	Dichlorofluoromethane	$CHCl_2F$	825.15	[43]
3	Dichloromethane	CH_2Cl_2	878	[70]
4	Formic acid	CH_2O_2	874.26	[43]
5	Methyl chloride	CH_3Cl	905	[43]
6	Methyl bromide	CH_3Br	810.37	[43]
7	Nitrocarbol	CH_3NO_2	652.15	[43]
8	Nitromethane	CH_3NO_2	643	[12]
9	Methanol	CH_4O	728	[70]
10	Hydroperoxymethane	CH_4O_2	387	[12]
11	Aminomethane	CH_5N	703.15	[43]
12	Trichloroethylene	C_2HCl_3	693	[116]
13	Ethyne	C_2H_2	578.15	[43]
14	Acetylene	C_2H_2	578	[70]
15	Acetylene tetrabromide	$C_2H_2Br_4$	608.15	[43]
16	*trans*-1,2-dichloroethylene	$C_2H_2Cl_2$	733	[43]
17	*cis*-1,2-Dichloroethylene	$C_2H_2Cl_2$	733	[43]
18	Vinylidene chloride	$C_2H_2Cl_2$	843	[43]
19	1,1,1-Trichloroethane	$C_2H_3Cl_3$	810.15	[43]
20	1,1,2-Trichloroethane	$C_2H_3Cl_3$	733.15	[43]
21	Acetonitrile	C_2H_3N	797	[70]
22	Vinyl chloride	C_2H_3Cl	745	[43]
23	Acetyl chloride	C_2H_3ClO	663.15	[43]
24	1,1-Difluoro-1-chloroethane	$C_2H_3ClF_2$	905	[43]
25	Ethylene	C_2H_4	723.15	[43]
26	Fluoroethene	C_2H_3F	658.15	[43]
27	Methyl formate	$C_2H_4O_2$	729.26	[43]
28	Ethylene oxide	C_2H_4O	702.04	[43]
29	Acetic acid	$C_2H_4O_2$	737	[116]
30	Methyl formate	$C_2H_4O_2$	729.26	[43]
31	1,1-Dichloroethane	$C_2H_4Cl_2$	731.15	[43]
32	1,2-Dichloroethane	$C_2H_4Cl_2$	711	[116]
33	Peroxyacetic acid	$C_2H_4O_3$	473.15	[43]
34	Acetaldehyde	C_2H_4O	758.15	[43]
35	Ethaneperoxoic acid (Peracetic acid)	$C_2H_4O_3$	383.15	[135]
36	Ethyl chloride	C_2H_5Cl	792	[43]
37	Nitroethane	$C_2H_5NO_2$	633.15	[43]
38	Ethylene chlorohydrin	C_2H_5ClO	698.15	[43]
39	Ethyl bromide	C_2H_5Br	784.26	[43]
40	Ethyleneimine	C_2H_5N	593.15	[43]

(continued)

No.	Compound	Formula	Exp.	Ref.
41	Nitroethane	$C_2H_5NO_2$	543.15	[12]
42	Ethyl nitrate	$C_2H_5NO_3$	358.15	[259]
43	Ethane	C_2H_6	745	[43]
44	Dimethyl ether	C_2H_6O	623.15	[43]
45	Ethanol	C_2H_6O	673	[116]
46	Ethylene glycol	$C_2H_6O_2$	673.15	[43]
47	Hydroperoxyethane	$C_2H_6O_2$	385	[12]
48	2-Aminoethanol	C_2H_7NO	673	[116]
49	Ethylamine	C_2H_7N	657	[43]
50	N,N-Dimethylamine	C_2H_7N	673.15	[43]
51	Ethanamine	C_2H_7N	657	[43]
52	Ethylenediamine	$C_2H_8N_2$	658.15	[43]
53	Urea peroxide	$C_2H_{10}N_4O_4$	633.15	[137]
54	Tetrafluorethene	C_2F_4	473.15	[43]
55	Acrylonitrile	C_3H_3N	754.26	[43]
56	Propyne	C_3H_4	613	[43]
57	2-Propenal	C_3H_4O	573	[116]
58	1-Propyne-3-ol	C_3H_4O	388.15	[43]
59	2-Propenoic acid	$C_3H_4O_2$	688	[70]
60	Acrylic acid	$C_3H_4O_2$	711.15	[43]
61	Allyl chloride	C_3H_5Cl	663	[70]
62	Propionitrile	C_3H_5N	785	[43]
63	Nitroglycerine	$C_3H_5N_3O_9$	543.15	[260]
64	3-Hydroxypropionitrile	C_3H_5NO	767.59	[43]
65	Propylene	C_3H_6	728.15	[43]
66	Cyclopropane	C_3H_6	770.93	[43]
67	Propene	C_3H_6	728	[70]
68	Allyl alcohol	C_3H_6O	643	[116]
69	Methyl vinyl ether	C_3H_6O	560.15	[43]
70	Propylene oxide	C_3H_6O	738.15	[43]
71	Propionaldehyde	C_3H_6O	500	[116]
72	Acetone	C_3H_6O	738	[70]
73	Propionic acid	$C_3H_6O_2$	713	[116]
74	Ethyl formate	$C_3H_6O_2$	708	[116]
75	Methyl acetate	$C_3H_6O_2$	748	[116]
76	1,3,5-Trioxacyclohexane	$C_3H_6O_3$	683	[70]
77	1,2-Dichloropropane	$C_3H_6Cl_2$	830	[43]
78	1,3,5-Trinitroperhydro-1,3,5-triazine (RDX)	$C_3H_6N_6O_6$	470.15	[261]
79	Allylamine	C_3H_7N	647.04	[43]
80	1-Bromopropane	C_3H_7Br	763.15	[43]
81	1-Chloropropane	C_3H_7Cl	793.15	[43]
82	2-Chloropropane	C_3H_7Cl	863	[70]
83	N,N-Dimethylformamide	C_3H_7NO	683	[70]
84	1-Nitropropane	$C_3H_7NO_2$	694.15	[43]

(continued)

(continued)

No.	Compound	Formula	Exp.	Ref.
85	2-Nitropropane	$C_3H_7NO_2$	578.15	[135]
86	1-Nitropropane	$C_3H_7NO_2$	643	[12]
87	2-Nitropropane	$C_3H_7NO_2$	698	[70]
88	Isopropyl nitrate	$C_3H_7NO_3$	448.15	[260]
89	n-Propyl nitrate	$C_3H_7NO_3$	448.15	[262]
90	1,3-Propylene glycol	$C_3H_8O_2$	651	[43]
91	Propane	C_3H_8	723	[43]
92	Ethyl methyl ether	C_3H_8O	463.15	[43]
93	1-Propanol	C_3H_8O	644.26	[43]
94	2-Propanol	C_3H_8O	672.04	[43]
95	Methylal	$C_3H_8O_2$	510.35	[43]
96	2-Methoxyethanol	$C_3H_8O_2$	558.15	[43]
97	1,2-Propanediol	$C_3H_8O_2$	694.26	[43]
98	Glycerol	$C_3H_8O_3$	673	[116]
99	Isopropylamine	C_3H_9N	675.15	[43]
100	Propylamine	C_3H_9N	591	[43]
101	DL -1-Amino-2-propanol	C_3H_9NO	647.04	[43]
102	1,2-Propanediamine	$C_3H_{10}N_2$	689	[116]
103	Maleic anhydride	$C_4H_2O_3$	749.82	[43]
104	Imidole	C_4H_5N	823	[70]
105	1,3-Butadiene	C_4H_6	702.04	[43]
106	Ethylacetylene	C_4H_6	592	[43]
107	Crotonaldehyde	C_4H_6O	553	[116]
108	2-Methyl-2-propenal	C_4H_6O	507.15	[43]
109	Vinyl ether	C_4H_6O	633.15	[43]
110	Vinyl acetate	$C_4H_6O_2$	698	[116]
111	Crotonic acid	$C_4H_6O_2$	669.26	[43]
112	Isocrotonic acid	$C_4H_6O_2$	669	[43]
113	Acetic acid anhydride	$C_4H_6O_3$	603	[70]
114	Diglycolic acid	$C_4H_6O_5$	503	[43]
115	Tartaric acid	$C_4H_6O_6$	700.93	[43]
116	Butane nitrile	C_4H_7N	761	[70]
117	2-Methylpropene	C_4H_8	738.15	[43]
118	*trans*-2-Butene	C_4H_8	597.04	[43]
119	Cyclobutane	C_4H_8	700	[43]
120	*cis*-2-Butene	C_4H_8	598.15	[43]
121	1-Butene	C_4H_8	657.04	[43]
122	Isobutylene	C_4H_8	738	[70]
123	2-Butanone	C_4H_8O	677	[116]
124	Butyraldehyde	C_4H_8O	503.15	[43]
125	Isobutyraldehyde	C_4H_8O	534	[116]
126	Ethyl vinyl ether	C_4H_8O	451	[70]
127	1,2-Epoxybutane	C_4H_8O	643	[70]
128	Tetrahydrofuran	C_4H_8O	594.26	[43]

(continued)

No.	Compound	Formula	Exp.	Ref.
129	Butyric acid	$C_4H_8O_2$	718	[116]
130	Ethyl acetate	$C_4H_8O_2$	700	[43]
131	Methyl propionate	$C_4H_8O_2$	728	[116]
132	Isobutyric acid	$C_4H_8O_2$	733	[116]
133	Propyl formate	$C_4H_8O_2$	708	[116]
134	1,4-Dioxane	$C_4H_8O_2$	453.15	[43]
135	Isopropyl Formate	$C_4H_8O_2$	713	[116]
136	Octahydro-1,3,5,7-tetranitro-1,3,5,7-tetrazocine (HMX)	$C_4H_8N_8O_8$	507.15	[261]
137	Tetrahydropyrrole	C_4H_9N	618	[70]
138	2-Hydroxy-1-ethylaziridine	C_4H_9NO	607	[70]
139	Morpholine	C_4H_9NO	583.15	[43]
140	N,N-Dimethylacetamide	C_4H_9NO	627.15	[43]
141	2-Butanone oxime	C_4H_9NO	588	[70]
142	1-Nitrobutane	$C_4H_9NO_2$	617	[12]
143	1-Chlorobutane	C_4H_9Cl	523	[116]
144	1-Bromobutane	C_4H_9Br	538	[70]
145	Butane	C_4H_{10}	645	[116]
146	2-Methylpropane	C_4H_{10}	733.15	[43]
147	Piperazine	$C_4H_{10}N_2$	728.15	[43]
148	1-Butanol	$C_4H_{10}O$	616	[43]
149	2-Butanol	$C_4H_{10}O$	663	[70]
150	tert-Butanol	$C_4H_{10}O$	733	[116]
151	2-Methyl-1-propanol	$C_4H_{10}O$	678	[116]
152	Ethoxy ethane	$C_4H_{10}O$	433.15	[43]
153	1,2-Dimethoxyethane,	$C_4H_{10}O_2$	473	[70]
154	2-Ethoxyethanol	$C_4H_{10}O_2$	508.15	[43]
155	1,4-Butanediol	$C_4H_{10}O_2$	630	[43]
156	1-Hydroperoxybutane	$C_4H_{10}O_2$	633.15	[12]
157	2-Hydroperoxy-2-methylpropane	$C_4H_{10}O_2$	360.15	[135]
158	1,3-Butanediol	$C_4H_{10}O_2$	667.04	[43]
159	2,2 -Dihydroxyethyl ether	$C_4H_{10}O_5$	502.04	[43]
160	1-Butanamine	$C_4H_{11}N$	585	[43]
161	Diethylamine	$C_4H_{11}N$	585.15	[43]
162	2-Methyl-1-propanamine	$C_4H_{11}N$	651.15	[43]
163	2-Butanamine	$C_4H_{11}N$	651	[43]
164	2-Aminoethylethanolamine	$C_4H_{12}N_2O$	641	[43]
165	Diethylenetriamine	$C_4H_{13}N_3$	631	[43]
166	Hexachlorobutadiene	C_4Cl_6	883.15	[43]
167	2-Furancarboxaldehyde	$C_5H_4O_2$	588.71	[43]
168	Azabenzene	C_5H_5N	823	[70]
169	Cyclopentadiene	C_5H_6	913.15	[43]
170	2-Methyl-1-buten-3-yne	C_5H_6	559	[43]
171	1-Penten-4-yne	C_5H_6	541	[43]

(continued)

(continued)

No.	Compound	Formula	Exp.	Ref.
172	1-Penten-3-yne	C_5H_6	559	[43]
173	2-Furanmethanol	$C_5H_6O_2$	664.15	[43]
174	1,3-Pentadiene	C_5H_8	613	[116]
175	Cyclopentene	C_5H_8	668.15	[43]
176	2-Methyl-1,3-butadiene	C_5H_8	493	[43]
177	1-Pentyne	C_5H_8	555	[43]
178	2-Pentyne	C_5H_8	575	[43]
179	Isoprene	C_5H_8	493.15	[43]
180	3-Methyll-1-butyne	C_5H_8O	568	[43]
181	Ethyl acrylate	$C_5H_8O_2$	655.93	[43]
182	Allyl acetate	$C_5H_8O_2$	647.04	[43]
183	2,4-Pentanedione	$C_5H_8O_2$	613.15	[43]
184	Methyl acetylacetate	$C_5H_8O_3$	623	[70]
185	Pentaerythritol tetranitrate	$C_5H_8N_4O_{12}$	433.15	[261]
186	1-Methyl-2-pyrrolidinone	C_5H_9NO	619.15	[43]
187	1-Pentene	C_5H_{10}	571	[116]
188	Cyclopentane	C_5H_{10}	593	[116]
189	trans-2-Pentene	C_5H_{10}	558	[43]
190	3-Methyl-1-butene	C_5H_{10}	638.15	[43]
191	2-Pentene	C_5H_{10}	561	[43]
192	2-Methyl-1-butene	C_5H_{10}	638	[43]
193	2-Methyl-2-butene	C_5H_{10}	563	[43]
194	Cis-2-Pentene	C_5H_{10}	561	[70]
195	Isopentene	C_5H_{10}	638	[70]
196	2-Methyl -2-butene	C_5H_{10}	638	[43]
197	2-Pentanone	$C_5H_{10}O$	725.15	[43]
198	3-Pentanone	$C_5H_{10}O$	725.37	[43]
199	Pentanal	$C_5H_{10}O$	495.15	[43]
200	Isopentanoic acid	$C_5H_{10}O_2$	689.15	[43]
201	Pentanoic acid	$C_5H_{10}O_2$	673.15	[43]
202	Propyl acetate	$C_5H_{10}O_2$	708	[116]
203	Ethyl propionate	$C_5H_{10}O_2$	718	[116]
204	Methyl butyrate	$C_5H_{10}O_2$	728	[116]
205	2,2-Dimethylpropionic acid	$C_5H_{10}O_2$	723	[116]
206	n-Butyl formate	$C_5H_{10}O_2$	595.37	[43]
207	Isopropyl Acetate	$C_5H_{10}O_2$	698	[116]
208	Tetrahydro-2-furancarbinol	$C_5H_{10}O_2$	555.37	[43]
209	Isobutyl formate	$C_5H_{10}O_2$	593.15	[43]
210	Ethyl 2-hydroxypropanoate	$C_5H_{10}O_3$	673.15	[43]
211	Amyl nitrite	$C_5H_{11}NO_2$	478	[70]
212	Isopentyl nitrite	$C_5H_{11}NO_2$	481	[70]
213	n-Pentyl nitrate (n-Amyl nitrate)	$C_5H_{11}NO_3$	468.15	[260]
214	1-Chloropentane	$C_5H_{11}Cl$	533.15	[43]
215	Pentane	C_5H_{12}	538	[116]

(continued)

No.	Compound	Formula	Exp.	Ref.
216	2-Methylbutane	C_5H_{12}	693.15	[43]
217	2,2-Dimethylpropane	C_5H_{12}	723.15	[43]
218	3-Pentanol	$C_5H_{12}O$	638	[116]
219	2-Methyl-1-butanol	$C_5H_{12}O$	658.15	[43]
220	3-Methyl-1-butanol	$C_5H_{12}O$	623.15	[43]
221	2-Pentanol	$C_5H_{12}O$	616.48	[43]
222	2-Methyl-2-butanol	$C_5H_{12}O$	708	[116]
223	2,2-Dimethyl-1-propanol	$C_5H_{12}O$	693	[116]
224	3-Methyl-2-butanol	$C_5H_{12}O$	710	[43]
225	2-Methoxy-2-methylpropane	$C_5H_{12}O$	708	[70]
226	1-Pentanol	$C_5H_{12}O$	573.15	[43]
227	2,2-Dimethyl-1,3-propanediol	$C_5H_{12}O_2$	672	[43]
228	1,5-Pentanediol	$C_5H_{12}O_2$	608.15	[43]
229	2-(2-Methoxyethoxy)ethanol	$C_5H_{12}O_3$	488	[70]
230	2,4,6-Trinitrophenol (Picric acid)	$C_6H_3N_3O_7$	573.15	[263]
231	1,3,5-Trichlorobenzene,	$C_6H_3Cl_3$	850	[43]
232	1,2,4-Trichlorobenzol	$C_6H_3Cl_3$	844.26	[43]
233	1,3-Dichloro-4-nitrobenzene	$C_6H_3Cl_2NO_2$	773.15	[70]
234	1,4-Dichloro-2-nitrobenzene	$C_6H_3Cl_2NO_2$	738.15	[70]
235	1-Chloro-2,3-dinitrobenzene	$C_6H_3ClN_2O_4$	705.15	[70]
236	1-Chloro-2,4-dinitrobenzene	$C_6H_3ClN_2O_4$	705.15	[12]
237	1,2-Dichloro-4-nitrobenzene	$C_6H_3Cl_2NO_2$	693.15	[70]
238	1-Chloro-3-nitrobenzene	$C_6H_4ClNO_2$	773.15	[70]
239	1-Chloro-2-nitrobenzene	$C_6H_4ClNO_2$	760.15	[70]
240	1-Chloro-4-nitrobenzene	$C_6H_4ClNO_2$	783.15	[70]
241	1,2-Dinitrobenzene	$C_6H_4N_2O_4$	743.15	[70]
242	1,4-Dichlorobenzene	$C_6H_4Cl_2$	920	[43]
243	1,2-Dichlorobenzene	$C_6H_4Cl_2$	913	[70]
244	1,3-Dichlorobenzene	$C_6H_4Cl_2$	920	[43]
245	Nitrobenzol	$C_6H_5NO_2$	753	[70]
246	Nitrobenzene	$C_6H_5NO_2$	578	[12]
247	4-Nitrophenol	$C_6H_5NO_3$	556.15	[263]
248	4-Hydroxynitrobenzene	$C_6H_5NO_3$	729	[70]
249	Chlorobenzene	C_6H_5Cl	863	[70]
250	Bromobenzene	C_6H_5Br	838.15	[43]
251	Benzene	C_6H_6	771	[70]
252	Phenol	C_6H_6O	878	[116]
253	*p*-Hydroquinone	$C_6H_6O_2$	788.7	[43]
254	*m*-Nitroaniline	$C_6H_6N_2O_2$	794	[70]
255	*o*-Nitroaniline	$C_6H_6N_2O_2$	794.15	[43]
256	*p*-Nitroaniline	$C_6H_6N_2O_2$	773	[70]
257	2-Nitroaniline	$C_6H_6N_2O_2$	794.15	[12]
258	3-Nitroaniline	$C_6H_6N_2O_2$	794.15	[135]
259	4-Nitroaniline	$C_6H_6N_2O_2$	783.15	[263]

(continued)

(continued)

No.	Compound	Formula	Exp.	Ref.
260	3-Picoline	C_6H_7N	810	[43]
261	Cyanoacetic ester		733	[70]
262	4-Picoline	C_6H_7N	810	[43]
263	Aniline	C_6H_7N	813	[2]
264	1,3-Cyclohexadiene	C_6H_8	633	[116]
265	1-Methyl-1,3-cyclopentadiene	C_6H_8	718	[43]
266	Hexanedinitrile	$C_6H_8N_2$	823.15	[43]
267	Cyclohexene	C_6H_{10}	583.15	[43]
268	1,3-Hexadiene	C_6H_{10}	593	[116]
269	1,5-Hexadiene	C_6H_{10}	618	[116]
270	1-Hexyne	C_6H_{10}	536	[70]
271	3-Hexyne	C_6H_{10}	554	[43]
272	2-Hexyne	C_6H_{10}	554	[43]
273	3-Methylpenta-1,4-diene	C_6H_{10}	586	[43]
274	(3E)-Hexa-1,3-diene	C_6H_{10}	587	[43]
275	Cyclohexanone	$C_6H_{10}O$	693.15	[43]
276	2-Ethylcrotonaldehyde	$C_6H_{10}O$	523	[116]
277	4-Methyl-3-penten-2-one	$C_6H_{10}O$	617.59	[43]
278	3-Hexyn-2,5-diol	$C_6H_{10}O_2$	553	[70]
279	Propyl acrylate	$C_6H_{10}O_2$	615	[43]
280	Propanoic acid anhydride	$C_6H_{10}O_3$	558	[43]
281	Ethyl acetylacetate	$C_6H_{10}O_3$	568.15	[43]
282	Hexanedioic acid	$C_6H_{10}O_4$	678	[70]
283	Ethylene diacetate	$C_6H_{10}O_4$	755.15	[43]
284	*trans*-2-Hexene	C_6H_{12}	528	[116]
285	4-Methyl-1-pentene	C_6H_{12}	577	[116]
286	2-Ethyl-1-butene	C_6H_{12}	597	[116]
287	2,3-Dimethyl-1-butene	C_6H_{12}	633.15	[43]
288	2-Methyl-1-pentene	C_6H_{12}	579	[116]
289	Methylcyclopentane	C_6H_{12}	602.04	[43]
290	1-Hexene	C_6H_{12}	538	[70]
291	2,3-Dimethyl-2-butene	C_6H_{12}	673.7	[43]
292	Cyclohexane	C_6H_{12}	533.15	[43]
293	*cis*-2-Hexene	C_6H_{12}	526	[43]
294	Ethylcyclobutane	C_6H_{12}	483	[43]
295	cis-3-Hexene	C_6H_{12}	553	[43]
296	trans-3-Hexene	C_6H_{12}	549	[70]
297	Isohexene	C_6H_{12}	577	[70]
298	3,3-Dimethyl L-1-butene	C_6H_{12}	630	[43]
299	3-Methyl -1-pentene	C_6H_{12}	625	[43]
300	Cyclohexanol	$C_6H_{12}O$	573.15	[43]
301	Butyl vinyl ether	$C_6H_{12}O$	528	[116]
302	2-Hexanone	$C_6H_{12}O$	697.04	[43]
303	4-Methyl-2-pentanone	$C_6H_{12}O$	721	[70]

(continued)

No.	Compound	Formula	Exp.	Ref.
304	Hexanoic acid	$C_6H_{12}O_2$	653.15	[43]
305	2-Methylpentanoic acid	$C_6H_{12}O_2$	651	[116]
306	Butyl acetate	$C_6H_{12}O_2$	653	[116]
307	Isobutyl acetate	$C_6H_{12}O_2$	696	[43]
308	*tert*-Butyl acetate	$C_6H_{12}O_2$	708	[116]
309	*sec*-Butyl acetate	$C_6H_{12}O_2$	683	[116]
310	*n*-Propyl propionate	$C_6H_{12}O_2$	703	[116]
311	Isopropyl propionate	$C_6H_{12}O_2$	698	[116]
312	Ethyl butyrate	$C_6H_{12}O_2$	713	[116]
313	4-Hydroxy-4-methyl-2-pentanone	$C_6H_{12}O_2$	876.48	[43]
314	2-Ethylbutyric acid	$C_6H_{12}O_2$	663	[116]
315	Triacetaldehyde	$C_6H_{12}O_3$	510.93	[43]
316	1-Methoxy-2-propyl acetate	$C_6H_{12}O_3$	627.15	[43]
317	2-Ethoxyethyl acetate	$C_6H_{12}O_3$	652.59	[43]
318	Hexahydro-1H-azepine	$C_6H_{13}N$	603.15	[43]
319	Cyclohexanamine	$C_6H_{13}N$	566.15	[43]
320	Hexane	C_6H_{14}	513	[116]
321	3-Methylpentane	C_6H_{14}	551.15	[43]
322	2,2-Dimethylbutane	C_6H_{14}	678	[70]
323	2-Methylpentane	C_6H_{14}	579.26	[43]
324	2,3-Dimethylbutane	C_6H_{14}	669	[116]
325	3-Methylpentane	C_6H_{14}	573	[43]
326	1-Hexanol	$C_6H_{14}O$	558	[43]
327	4-Methyl-2-pentanol	$C_6H_{14}O$	613	[116]
328	Dipropyl ether	$C_6H_{14}O$	488	[70]
329	Diisopropyl ether	$C_6H_{14}O$	678	[70]
330	1,1-Diethoxyethane	$C_6H_{14}O_2$	503.15	[43]
331	2-Butoxyethanol	$C_6H_{14}O_2$	511.15	[43]
332	2,4-Dihydroxy-2-methylpentane	$C_6H_{14}O_2$	698	[70]
333	2-Methoxyethyl ether	$C_6H_{14}O_3$	463	[70]
334	2-(2-Ethoxyethoxy) ethanol	$C_6H_{14}O_3$	477.15	[43]
335	Triethylene glycol	$C_6H_{14}O_4$	644	[43]
336	Dipropylamine	$C_6H_{15}N$	572.15	[43]
337	Diisopropylamine	$C_6H_{15}N$	588.71	[43]
338	2-Diethylaminoethanol	$C_6H_{15}NO$	593	[116]
339	Diisopropanolamine	$C_6H_{15}NO_2$	647	[116]
340	2,4,6-Trinitrotoluene (TNT)	$C_7H_5N_3O_6$	548.15	[261]
341	N-Methyl-N,2,4,6-tetranitroaniline (Tetryl)	$C_7H_5N_5O_8$	439.15	[261]
342	Benzoyl chloride	C_7H_5ClO	873	[70]
343	Benzoic acid	$C_7H_6O_2$	805	[70]
344	2-Hydroxybenzoic acid	$C_7H_6O_3$	818.15	[43]
345	2-Methyl-3,4-dinitrophenol	$C_7H_6N_2O_5$	613.15	[70]
346	2-Methyl-4,6-dinitrophenol	$C_7H_6N_2O_5$	613.15	[135]
347	4-Fluorobenzyl chloride	C_7H_6ClF	863	[70]

(continued)

(continued)

No.	Compound	Formula	Exp.	Ref.
348	2-Methylnitrobenzene	$C_7H_7NO_2$	693	[70]
349	1-Methyl-3-nitrobenzene	$C_7H_7NO_2$	713	[70]
350	1-Methyl-2-nitrobenzene	$C_7H_7NO_2$	578.15	[135]
351	1-Methyl-3-nitrobenzene	$C_7H_7NO_2$	578	[12]
352	1-Methyl-4-nitrobenzene	$C_7H_7NO_2$	616	[12]
353	(Nitromethyl) benzene	$C_7H_7NO_2$	578.15	[135]
354	Benzyl chloride	C_7H_7Cl	858.15	[43]
355	Toluene	C_7H_8	755	[116]
356	Benzyl alcohol	C_7H_8O	709.26	[43]
357	Methoxybenzene	C_7H_8O	748	[70]
358	*m*-Cresol	C_7H_8O	832.04	[43]
359	*o*-Cresol	C_7H_8O	872.04	[43]
360	*p*-Cresol	C_7H_8O	832.04	[43]
361	3-Aminotoluene	C_7H_9N	755.15	[43]
362	4-Aminotoluene	C_7H_9N	755.15	[43]
363	2-Aminotoluene	C_7H_9N	755.15	[43]
364	2-Norbornene	C_7H_{10}	778	[70]
365	Isobutyl acrylate	$C_7H_{12}O_2$	613.15	[43]
366	*n*-Butyl acrylate	$C_7H_{12}O_2$	565.93	[43]
367	1-Heptene	C_7H_{14}	536	[116]
368	2,3,3-Trimethyl-1-butene	C_7H_{14}	656	[116]
369	Ethylcyclopentane	C_7H_{14}	533.5	[43]
370	Methylcyclohexane	C_7H_{14}	558.15	[43]
371	Cycloheptane	C_7H_{14}	428	[43]
372	2,3,3-Trimethyl-1butene	C_7H_{14}	656	[116]
373	cis-2-Heptene	C_7H_{14}	512	[43]
374	trans-2-Heptene	C_7H_{14}	517	[43]
375	cis-3-Heptene	C_7H_{14}	563	[43]
376	trans-3-Heptene	C_7H_{14}	577	[43]
377	2-Methyl -1-hexene	C_7H_{14}	543	[43]
378	2-Ethyl-1-pentene	C_7H_{14}	600	[43]
379	3-Ethyl -1-pentene	C_7H_{14}	600	[43]
380	Methylhexanone	$C_7H_{14}O$	728	[70]
381	*cis*-2-Methylcyclohexanol	$C_7H_{14}O$	569.15	[43]
382	*cis*-4-Methylcyclohexanol	$C_7H_{14}O$	570.15	[43]
383	2-Heptanone	$C_7H_{14}O$	666.15	[43]
384	*trans*-2-Methylcyclohexanol	$C_7H_{14}O$	569.15	[43]
385	*trans*-4-Methylcyclohexanol	$C_7H_{14}O$	570.15	[43]
386	Heptanoic acid	$C_7H_{14}O_2$	571	[43]
387	Pentyl acetate	$C_7H_{14}O_2$	633.15	[43]
388	Isopentyl Acetate	$C_7H_{14}O_2$	653	[116]
389	Butyl propionate	$C_7H_{14}O_2$	658	[116]
390	Isobutyl propionate	$C_7H_{14}O_2$	708	[116]
391	Propyl butyrate	$C_7H_{14}O_2$	693	[116]

(continued)

No.	Compound	Formula	Exp.	Ref.
392	Isopropyl butyrate	$C_7H_{14}O_2$	708	[43]
393	Heptane	C_7H_{16}	486	[116]
394	2,3-Dimethylpentane	C_7H_{16}	610.37	[43]
395	2,2,3-Trimethylbutane	C_7H_{16}	685	[116]
396	2,2,4-Trimethylbutane	C_7H_{16}	680	[116]
397	2,4-Dimethylpentane	C_7H_{16}	610	[43]
398	3-Methylhexane	C_7H_{16}	553.15	[43]
399	3,3-Dimethylpentane	C_7H_{16}	610	[43]
400	2-Methylhexane	C_7H_{16}	566	[70]
401	3,3-Dimethylpentane	C_7H_{16}	610	[70]
402	2,4-Dimethylpentane	C_7H_{16}	610	[70]
403	2,3-Dimethylpentane	C_7H_{16}	610	[70]
404	3-Methylhexane	C_7H_{16}	553	[70]
405	1-Heptanol	$C_7H_{16}O$	555	[43]
406	4-Heptanol	$C_7H_{16}O$	568	[116]
407	2,4-Dimethyl-3-pentanol	$C_7H_{16}O$	668	[116]
408	Phthalic anhydride	$C_8H_4O_3$	857.04	[43]
409	Phenylacetylene	C_8H_6	763	[43]
410	Ethynylbenzene	C_8H_6	763	[43]
411	1,4-Benzenedicarboxylic acid	$C_8H_6O_4$	769	[43]
412	1,3-Benzenedicarboxylic acid	$C_8H_6O_4$	769	[43]
413	o-Phthalic acid	$C_8H_6O_4$	863	[116]
414	Styrene	C_8H_8	763.15	[43]
415	Acetophenone	C_8H_8O	833	[116]
416	Phenyl acetate	$C_8H_8O_2$	858	[116]
417	Methyl benzoate	$C_8H_8O_2$	783	[70]
418	2-Hydroxybenzoic acid methyl ester	$C_8H_8O_3$	728	[43]
419	N-Phenylacetamide	C_8H_9NO	803.15	[43]
420	1,3-Dimethyl-5-nitrobenzene	$C_8H_9NO_2$	647	[12]
421	Ethylbenzene	C_8H_{10}	705.37	[43]
422	1,3-Dimethylbenzene	C_8H_{10}	800.93	[43]
423	1,2-Dimethylbenzene	C_8H_{10}	737.04	[43]
424	1,4-Dimethylbenzene	C_8H_{10}	802.04	[43]
425	P-Xylene	C_8H_{10}	802	[43]
426	O-Xylene	C_8H_{10}	737	[43]
427	m-Xylene	C_8H_{10}	738	[43]
428	3,5-Dimethylphenol	$C_8H_{10}O$	828	[116]
429	2,4-Dimethylphenol	$C_8H_{10}O$	872	[43]
430	2,6-Xylenol	$C_8H_{10}O$	872.04	[43]
431	2,3-Dimethylphenol	$C_8H_{10}O$	872	[43]
432	3,4-Dimethylphenol	$C_8H_{10}O$	872.04	[43]
433	2,5-Dimethylphenol	$C_8H_{10}O$	872	[43]
434	Ethylphenylamine	$C_8H_{11}N$	752.15	[43]
435	*N,N*-Dimethylbenzenamine	$C_8H_{11}N$	644.26	[43]

(continued)

(continued)

No.	Compound	Formula	Exp.	Ref.
436	Vinylcyclohexene	C_8H_{12}	543	[43]
437	Cyclohexenylethylene	C_8H_{12}	543	[43]
438	1,7-Octadiene	C_8H_{14}	493	[70]
439	Hexylacetylene	C_8H_{14}	498	[70]
440	1-Octyne	C_8H_{14}	517	[43]
441	cis-2-Octene	C_8H_{14}	507	[43]
442	trans-2-Octene	C_8H_{14}	517	[43]
443	cis-3-Octene	C_8H_{14}	538	[43]
444	trans-3-Octene	C_8H_{14}	548	[43]
445	cis-4-Octene	C_8H_{14}	539	[43]
446	6-Methyl -1-heptene	C_8H_{14}	547	[43]
447	2-Ethyl-1-hexene	C_8H_{14}	600	[43]
448	2,4,4-Trimethylpentene	C_8H_{14}	693	[70]
449	3-Ethylhex-1-ene	C_8H_{14}	552	[43]
450	2,3-Dimethyl hexene	C_8H_{14}	590	[43]
451	4-Methylhept-1-ene	C_8H_{14}	548	[43]
452	Vinylcyclohexane	C_8H_{14}	543	[43]
453	Butanoic acid anhydride	$C_8H_{14}O_3$	552.59	[43]
454	1-Octene	C_8H_{16}	523	[70]
455	2,4,4-Trimethyl-1-pentene	C_8H_{16}	693	[43]
456	Ethylcyclohexane	C_8H_{16}	535.37	[43]
457	trans-1,2-Dimethylcyclohexane	C_8H_{16}	577.15	[43]
458	2,4,4-Trimethyl-2-pentene	C_8H_{16}	581	[43]
459	Propylcyclopentane	C_8H_{16}	542.15	[43]
460	trans-1,3-Dimethylcyclohexane	C_8H_{16}	579	[43]
461	trans-1,4-Dimethylcyclohexane	C_8H_{16}	577	[43]
462	1,1-Dimethylcyclohexane	C_8H_{16}	577	[43]
463	2,3,3-Trimethyl-1-Pentene	C_8H_{16}	504	[70]
464	cis-1,4-Dimethylcyclohexane	C_8H_{16}	577	[43]
465	cis-1,2-Dimethylcyclohexane	C_8H_{16}	577.15	[43]
466	cis-1,3-Dimethylcyclohexane	C_8H_{16}	579	[43]
467	3,4,4-Trimethyl-2-pentene	C_8H_{16}	598	[43]
468	Cyclooctane	C_8H_{16}	430	[43]
469	n-Propylcyclopentane	C_8H_{16}	542	[70]
470	1,1-Dmethylcyclohexane	C_8H_{16}	577	[70]
471	Cis-1,2-Dmethylcyclohexane	C_8H_{16}	577	[70]
472	trans-1,2-Dmethylcyclohexane	C_8H_{16}	577	[70]
473	Cis-1,3-Dmethylcyclohexane	C_8H_{16}	579	[70]
474	Trans-1,3-Dmethylcyclohexane	C_8H_{16}	579	[43]
475	2-Ethylhexanal	$C_8H_{16}O$	463.15	[43]
476	Isobutyl isobutyrate	$C_8H_{16}O_2$	705.15	[43]
477	Hexyl acetate	$C_8H_{16}O_2$	528	[116]
478	Butyl butyrate	$C_8H_{16}O_2$	623	[116]
479	Octanoic acid	$C_8H_{16}O_2$	570	[43]

(continued)

No.	Compound	Formula	Exp.	Ref.
480	Isopentyl propionate	$C_8H_{16}O_2$	698	[43]
481	2-(2-Ethoxyethoxy)ethyl acetate	$C_8H_{16}O_4$	583	[70]
482	Octane	C_8H_{18}	483	[70]
483	2,3,4-Trimethylpentane	C_8H_{18}	700	[70]
484	3,4-Dimethylhexane	C_8H_{18}	588	[70]
485	2,2,4-Trimethylpentane	C_8H_{18}	683	[70]
486	3,3-Dimethylhexane	C_8H_{18}	610	[70]
487	2,3-Dimethylhexane	C_8H_{18}	588	[70]
488	2,4-Dimethylhexane	C_8H_{18}	588	[70]
489	2,5-Dimethylhexane	C_8H_{18}	588	[70]
490	2,3,3-Trimethylpentane	C_8H_{18}	703	[70]
491	2-Methylheptane	C_8H_{18}	520	[70]
492	3-Methylheptane	C_8H_{18}	518	[70]
493	2,2-Dimethylhexane	C_8H_{18}	610	[70]
494	Dibutyl ether	$C_8H_{18}O$	467.59	[43]
495	2-Octanol	$C_8H_{18}O$	538	[116]
496	2-Ethyl-1-hexanol	$C_8H_{18}O$	560.93	[43]
497	1-Octanol	$C_8H_{18}O$	555	[43]
498	2-Ethyl-1,3-hexanediol	$C_8H_{18}O_2$	633	[116]
499	*n*-Hexyl Cellosolve	$C_8H_{18}O_2$	553	[116]
500	Bis(2-ethoxyethyl)ether	$C_8H_{18}O_3$	478	[116]
501	2-(2-Butoxyethoxy)ethanol	$C_8H_{18}O_3$	477.59	[43]
502	Methyl ethyl ketone peroxide	$C_8H_{18}O_6$	663.15	[260]
503	1-Octanamine	$C_8H_{19}N$	538	[70]
504	*N*-Ethyl-*N,N*-diisopropylamine	$C_8H_{19}N$	513	[70]
505	2-Amino-2-ethylhexane	$C_8H_{19}N$	538	[70]
506	N-Butyl-1-butanamine	$C_8H_{19}N$	533	[70]
507	Tetraethylenepentamine	$C_8H_{23}N_5$	594	[70]
508	1-Benzazine	C_9H_7N	753.15	[43]
509	Hydrindane	C_9H_{10}	569	[116]
510	Methylstyrene	C_9H_{10}	847.59	[43]
511	*cis*-1-Propenylbenzene	C_9H_{10}	848	[43]
512	2-Vinyltoluene	C_9H_{10}	767	[70]
513	3-vinyltoluene	C_9H_{10}	762	[70]
514	4-vinyltoluene	C_9H_{10}	848	[70]
515	Isopropenylbenzene	C_9H_{10}	718	[70]
516	trans-1-Propenylbenzene	C_9H_{10}	848	[43]
517	Ethyl benzoate	$C_9H_{10}O_2$	763.15	[43]
518	Benzyl acetate	$C_9H_{10}O_2$	734	[43]
519	*trans*-1-Methylstyrene	C_9H_{11}	848	[43]
520	*n*-Propylbenzene	C_9H_{12}	729.15	[43]
521	1,2,3-Trimethylbenezene	C_9H_{12}	743.15	[43]
522	1,2,4-Trimethylbenezene	C_9H_{12}	788.15	[43]
523	1-Methyl-2-ethylbenzene	C_9H_{12}	721	[116]

(continued)

(continued)

No.	Compound	Formula	Exp.	Ref.
524	1-Methyl-3-ethylbenzene	C_9H_{12}	753.15	[43]
525	1-Methyl-4-ethylbenzene	C_9H_{12}	748.15	[43]
526	Isopropylbenzene	C_9H_{12}	697.04	[43]
527	1,3,5-Trimethylbenzene	C_9H_{12}	823.15	[43]
528	Cumene	C_9H_{12}	693	[70]
529	m-Ethyltoluene	C_9H_{12}	753	[70]
530	1,2,3-Trimethylbenzene	C_9H_{12}	743	[70]
531	1,2,4-Trimethylbenzene	C_9H_{12}	788	[70]
532	o-Ethyltoluene	C_9H_{12}	713	[70]
533	p-Ethyltoluene	C_9H_{12}	808	[70]
534	m-DiEthylbenzene	C_9H_{12}	723	[70]
535	1-Hydroperoxy-2-isopropylbenzene	$C_9H_{12}O_2$	422.039	[264]
536	1-Hydroperoxy-2-propylbenzene	$C_9H_{12}O_2$	515	[12]
537	3,5,5-Trimethyl-2-cyclohexane-1-one	$C_9H_{14}O$	733.15	[43]
538	Glyceryl triacetate	$C_9H_{14}O_6$	706	[43]
539	n-Propylcyclohexane	C_9H_{18}	521.15	[43]
540	Isopropylcyclohexane	C_9H_{18}	556	[70]
541	1-Nonene	C_9H_{18}	510	[43]
542	Butylcyclopentane	C_9H_{18}	523.15	[43]
543	1,3,5-Trimethylcyclohexane	C_9H_{18}	587	[116]
544	Nonanoic acid	$C_9H_{18}O_2$	589	[43]
545	Nonane	C_9H_{20}	485	[70]
546	4-Methyloctane	C_9H_{20}	493	[70]
547	2-Methyloctane	C_9H_{20}	493	[70]
548	3-Methyloctane	C_9H_{20}	493	[70]
549	2,2,3,4-Tetramethylpentane	C_9H_{20}	703	[70]
550	2,2,4,4-Tetramethylpentane	C_9H_{20}	703	[70]
551	2,3,3,4-Tetramethylpentane	C_9H_{20}	703	[43]
552	Tetraethylmethane	C_9H_{20}	563	[70]
553	2,4-Dimethyl-3-ethylpentane	C_9H_{20}	663	[70]
554	2,2,5-Trimethylhexane	C_9H_{20}	623	[43]
555	2,2-Dimethylheptane	C_9H_{20}	588	[43]
556	1-Nonanol	$C_9H_{20}O$	533	[116]
557	1-Nonanol	$C_9H_{20}O$	550	[43]
558	2,3,3,3-Tetramethylpentane	C_9H_{21}	703	[43]
559	Triisopropanolamine	$C_9H_{21}NO_3$	593	[116]
560	Naphthalene	$C_{10}H_8$	813	[116]
561	Dimethyl terephthalate	$C_{10}H_{10}O_4$	843.15	[43]
562	1,2-Dimethyl phthalate	$C_{10}H_{10}O_4$	829	[43]
563	N-Phenylacetoacetamide	$C_{10}H_{11}NO_2$	725.15	[43]
564	Tetralin	$C_{10}H_{12}$	657.04	[43]
565	Dicyclopentadiene	$C_{10}H_{12}$	783.15	[43]
566	2-Ethyl-1-vinylbenzene	$C_{10}H_{12}$	634	[43]
567	3-Ethyl -1-vinylbenzene	$C_{10}H_{12}$	657	[43]

(continued)

No.	Compound	Formula	Exp.	Ref.
568	2-(sec-Butyl)-4,6-dinitrophenol	$C_{10}H_{12}N_2O_5$	616	[12]
569	n-Butylbenzene	$C_{10}H_{14}$	685.37	[43]
570	1,4-Diethylbenzene	$C_{10}H_{14}$	703.15	[43]
571	Isobutylbenzene	$C_{10}H_{14}$	700.93	[43]
572	sec-Butylbenzene	$C_{10}H_{14}$	690.93	[43]
573	tert-Butylbenzene	$C_{10}H_{14}$	723.15	[43]
574	1,2-Diethylbenzene	$C_{10}H_{14}$	677	[116]
575	1,3-Diethylbenzene	$C_{10}H_{14}$	723.15	[43]
576	3-Isopropyltoluene	$C_{10}H_{14}$	709	[43]
577	2-Isopropyltoluene	$C_{10}H_{14}$	650	[43]
578	p-Cymene	$C_{10}H_{14}$	709.26	[43]
579	1,1-Diphenylethane	$C_{14}H_{14}$	713.15	[43]
580	1,2,3,4-Tetramethylbenzene	$C_{10}H_{14}$	700	[43]
581	O-DiEthylbenzene	$C_{10}H_{14}$	668	[70]
582	Butylbenzene	$C_{10}H_{14}$	685	[43]
583	1,2,3,5-Tetramethylbenzene	$C_{10}H_{14}$	700	[70]
584	o-Cymene	$C_{10}H_{14}$	709	[70]
585	m-Cymene	$C_{10}H_{14}$	709	[70]
586	p-Diethylbenzene	$C_{10}H_{14}$	703	[70]
587	1-Butyl-2-hydroperoxybenzene	$C_{10}H_{14}O_2$	504	[12]
588	α-Pinene	$C_{10}H_{16}$	528.15	[43]
589	beta-Pinene	$C_{10}H_{16}$	528	[43]
590	Adamantane	$C_{10}H_{16}$	560	[43]
591	Decalin	$C_{10}H_{18}$	541	[116]
592	trans-Decahydronaphthalene	$C_{10}H_{18}$	528	[70]
593	cis-Decahydronaphthalene	$C_{10}H_{18}$	523.15	[43]
594	Deca-1,9-diene	$C_{10}H_{18}$	565	[43]
595	Dec-1-yne	$C_{10}H_{18}$	497	[43]
596	1-Decene	$C_{10}H_{20}$	508.15	[43]
597	Butylcyclohexane	$C_{10}H_{20}$	519.15	[43]
598	Isobutylcyclohexane	$C_{10}H_{20}$	547	[116]
599	sec-Butylcyclohexane	$C_{10}H_{20}$	550	[116]
600	tert-Butylcyclohexane	$C_{10}H_{20}$	615	[116]
601	4-Isopropyl-1-methylcyclohexane	$C_{10}H_{20}$	579	[116]
602	Cyclodecane	$C_{10}H_{20}$	508	[116]
603	n-Butylcyclohexane	$C_{10}H_{20}$	519	[43]
604	1,2,3,4-Tetramethylcyclohexane	$C_{10}H_{20}$	643	[43]
605	1,1-Diethylcyclohexane	$C_{10}H_{20}$	577	[43]
606	trans-1,4-Diethylcyclohexane	$C_{10}H_{20}$	577	[43]
607	Decanoic acid	$C_{10}H_{20}O_2$	570	[43]
608	Decane	$C_{10}H_{22}$	478	[70]
609	3-Methylnonane	$C_{10}H_{22}$	487	[70]
610	2-Methylnonane	$C_{10}H_{22}$	487	[43]
611	4-Methylnonane	$C_{10}H_{22}$	485	[43]

(continued)

(continued)

No.	Compound	Formula	Exp.	Ref.
612	5-Methylnonane	$C_{10}H_{22}$	485	[43]
613	3,3,5-Trimethylheptane	$C_{10}H_{22}$	655	[43]
614	2,2,3,3-Tetramethylhexane	$C_{10}H_{22}$	714	[43]
615	2,2,5,5-Tetramethylhexane	$C_{10}H_{22}$	752	[43]
616	2,2-Dimethyloctane	$C_{10}H_{22}$	571	[43]
617	1-Decanol	$C_{10}H_{22}O$	523	[116]
618	Dipentyl ether	$C_{10}H_{22}O$	444	[70]
619	1-Methylnaphthalene	$C_{11}H_{10}$	802.04	[43]
620	2-Methylnaphthalene	$C_{11}H_{10}$	802	[43]
621	1-methylnaphtalene	$C_{11}H_{10}$	802	[70]
622	2-methylnaphtalene	$C_{11}H_{10}$	761	[70]
623	Butyl benzoate	$C_{11}H_{14}O_2$	708	[116]
624	1-Methyl-3,5-diethylbenzene	$C_{11}H_{16}$	734	[116]
625	p-tert-Butyltoluene	$C_{11}H_{16}$	783	[70]
626	2,3-Dimethyl-1-propylbenzene	$C_{11}H_{16}$	738	[43]
627	1,2,3,4,5-Pentamethylbenzene	$C_{11}H_{16}$	703	[43]
628	1,2,3-Trimethyl-4-ethylbenzne	$C_{11}H_{16}$	766	[43]
629	1,2,4-Trimethyl-3-ethylbenzene	$C_{11}H_{16}$	766	[43]
630	1,2,4-Trimethyl -5-ethylbenzene	$C_{11}H_{16}$	766	[43]
631	2-Ethylhexyl acrylate	$C_{11}H_{20}O_2$	530.93	[43]
632	n-Hexylcyclopentane	$C_{11}H_{22}$	501	[116]
633	1-Hendecene	$C_{11}H_{22}$	510	[43]
634	1-Undecene	$C_{11}H_{22}$	510	[70]
635	Undecane	$C_{11}H_{24}$	513	[70]
636	1-Hendecanol	$C_{11}H_{24}O$	550	[43]
637	2-Nitro-1,1′-biphenyl	$C_{12}H_9NO_2$	452.15	[259]
638	Biphenyl	$C_{12}H_{10}$	813.15	[43]
639	Diphenyl ether	$C_{12}H_{10}O$	891.15	[43]
640	Diphenylamine	$C_{12}H_{11}N$	907.04	[43]
641	2-Aminobiphenyl	$C_{12}H_{11}N$	725	[116]
642	1-Ethylnaphthalene	$C_{12}H_{12}$	754	[116]
643	1-Ethylnaphtalene	$C_{12}H_{12}$	754	[21]
644	2-Ethylnaphthalene	$C_{12}H_{12}$	750	[43]
645	Diethyl phthalate	$C_{12}H_{14}O_4$	730.15	[43]
646	Nitrocellulose	$C_{12}H_{14}N_6O_{22}$	443.15	[265]
647	Cyclohexylbenzene	$C_{12}H_{16}$	813	[43]
648	1,2,4-Triethenyl-Cyclohexane	$C_{12}H_{18}$	543	[70]
649	1,3-Diisopropylbenzene	$C_{12}H_{18}$	722	[70]
650	3,5-Dimethyl-tert-butylbenzene	$C_{12}H_{18}$	833	[70]
651	Diisopropylbenzene	$C_{12}H_{18}$	721	[70]
652	m-Diisopropylbenzene	$C_{12}H_{18}$	722	[70]
653	p-tert-Butyl ethylbenzene	$C_{12}H_{18}$	553	[43]
654	1,2,4-Triethylbenzene	$C_{12}H_{18}$	640	[43]
655	1,2,3-Triethylbenzene	$C_{12}H_{18}$	640	[43]

(continued)

No.	Compound	Formula	Exp.	Ref.
656	1,3-Dimethyl adamantane	$C_{12}H_{20}$	615	[43]
657	Dicyclohexyl	$C_{12}H_{22}$	518.15	[43]
658	Bicyclohexyl	$C_{12}H_{22}$	518	[70]
659	Cyclohexanone peroxide	$C_{12}H_{22}O_6$	675.927	[137]
660	1-Dodecene	$C_{12}H_{24}$	528.15	[43]
661	Dodecanoic acid	$C_{12}H_{24}O_2$	503	[116]
662	n-Decyl acetate	$C_{12}H_{24}O_2$	488	[116]
663	3-Methylundecane	$C_{12}H_{26}$	480	[43]
664	Dodecane	$C_{12}H_{26}$	473	[70]
665	1-Dodecanol	$C_{12}H_{26}O$	548.15	[43]
666	Dihexyl ether	$C_{12}H_{26}O$	458.15	[43]
667	Phenyl benzoate	$C_{13}H_{10}O_2$	833	[43]
668	Diphenylmethane	$C_{13}H_{12}$	759	[116]
669	2-Methylbiphenyl	$C_{13}H_{12}$	775	[116]
670	1-Tridecene	$C_{13}H_{26}$	510	[70]
671	Tridecane	$C_{13}H_{28}$	475	[70]
672	Anthracene	$C_{14}H_{10}$	828	[116]
673	Benzoic peroxyanhydride (or Benzoyl peroxide)	$C_{14}H_{10}O_4$	353.15	[266]
674	2-Ethylbiphenyl	$C_{14}H_{14}$	722	[116]
675	1,2-Diphenylethane	$C_{14}H_{14}$	753	[43]
676	1-Butylnaphthalene	$C_{14}H_{16}$	633	[70]
677	1,4-DI-tert-butylbenzene	$C_{14}H_{22}$	496	[43]
678	1,2,3,5-Tetraethylbenzene	$C_{14}H_{22}$	598	[43]
679	1-Tetradecene	$C_{14}H_{28}$	508	[70]
680	Tetradecanoic acid	$C_{14}H_{28}O_2$	508	[116]
681	Tetradecane	$C_{14}H_{30}$	493	[70]
682	2-Propylbiphenyl	$C_{15}H_{16}$	725	[116]
683	1-Pentadecene	$C_{15}H_{30}$	510	[70]
684	Pentadecane	$C_{15}H_{32}$	475	[70]
685	Chloroprene	$C_{16}H_9Cl$	593.15	[43]
686	2-Butylbiphenyl	$C_{16}H_{18}$	706	[116]
687	1-Hexadecene	$C_{16}H_{32}$	513.15	[43]
688	Dibutyl phthalate	$C_{16}H_{22}O_4$	675.15	[43]
689	Pentaethylbenzene	$C_{16}H_{26}$	566	[43]
690	Hexadecanoic acid	$C_{16}H_{32}O_2$	413	[116]
691	1-Heptadecene	$C_{17}H_{34}$	510	[70]
692	Heptadecane	$C_{17}H_{36}$	475	[70]
693	Naphthacene	$C_{18}H_{12}$	830	[43]
694	p-Terphenyl	$C_{18}H_{14}$	808	[43]
695	Hexaethylbenzene	$C_{18}H_{30}$	541	[43]
696	2,4-Dicyclohexyl-2-methylpentane	$C_{18}H_{34}$	498	[70]
697	Dibutyl sebacate	$C_{18}H_{34}O_4$	638.15	[43]
698	1-Octadecene	$C_{18}H_{36}$	523	[70]

(continued)

(continued)

No.	Compound	Formula	Exp.	Ref.
699	Octadecane	$C_{18}H_{38}$	508	[70]
700	1-Nonylnaphthalene	$C_{19}H_{26}$	576	[43]
701	1-Nonadecene	$C_{19}H_{38}$	510	[70]
702	Nonadecae	$C_{19}H_{40}$	478	[70]
703	1-Decylnaphthalene	$C_{20}H_{28}$	567	[43]
704	1-Eicosene	$C_{20}H_{40}$	510	[70]
705	(2E)Icos-2-ene	$C_{20}H_{40}$	499	[43]
706	Eicosane	$C_{20}H_{42}$	475	[70]
707	Henicosane	$C_{21}H_{44}$	475	[70]
708	Tricosane	$C_{23}H_{48}$	475	[43]
709	Dodecanoic peroxyanhydride (Dilauroyl peroxide)	$C_{24}H_{46}O_4$	279.15	[260]
710	n-Tetracosane	$C_{24}H_{50}$	475	[43]
711	Pentacosane	$C_{25}H_{52}$	475	[43]
712	Hexacosane	$C_{26}H_{54}$	475	[43]
713	Heptacosane	$C_{27}H_{56}$	475	[43]
714	n-Octacosane	$C_{28}H_{58}$	475	[43]
715	Nonacosane	$C_{29}H_{60}$	475	[43]

Appendix C

The reported standard net heat of combustion (kJ mol^{-1}) of combustible organic compounds [59]

No.	Compound	Formula	ΔH_c°(net)
1	Methane	CH_4	800.18
2	Ethane	C_2H_6	1,428.65
3	Acetylene (ethyne)	C_2H_2	1,256.44
4	Propane	C_3H_8	2,043.12
5	Propylene (propene)	C_3H_6	1,985.07
6	Cyclopropane	C_3H_6	1,959.32
7	Propadiene	C_3H_4	1,856.23
8	Methylacetylene (propyne)	C_3H_4	1,848.50
9	Vinylacetylene	C_4H_4	2,340.27
10	Ethylene (ethene)	C_4H_6	1,323.22
11	1,2-butadiene	C_4H_6	2,461.79
12	1,3-butadiene	C_4H_6	2,409.70
13	Dimethylacetylene	C_4H_6	2,418.89
14	Ethylacetylene (1-butyne)	C_4H_6	2,464.68
15	Methylcyclopropane	C_4H_8	2,565.47
16	Cyclobutane	C_4H_8	2,567.95
17	1-Butene (butylene)	C_4H_8	2,540.80
18	Cis-2-Butene	C_4H_8	2,533.89
19	Trans-2-Butene	C_4H_8	2,530.49
20	2-methylpropene (Isobutene)	C_4H_8	2,524.23
21	N-Butane	C_4H_{10}	2,657.37
22	Isobutane	C_4H_{10}	2,648.18
23	N-Pentane	C_5H_{12}	3,271.67
24	Isopentane (2-methylbutane)	C_5H_{12}	3,239.61
25	Neopentane (2,2-dimethylpropane)	C_5H_{12}	3,250.52
26	Ethylcyclopropane	C_5H_{10}	3,180.07
27	Cis-1,2-Dimethylcyclopropane	C_5H_{10}	3,171.10
28	Methylcyclobutane	C_5H_{10}	3,173.87
29	Cyclopentane	C_5H_{10}	3,070.78
30	1-pentene	C_5H_{10}	3,129.83
31	Cis-2-Pentene	C_5H_{10}	3,123.14
32	Trans-2-Pentene	C_5H_{10}	3,118.74
33	2-Methyl-1-butene	C_5H_{10}	3,115.80
34	3-Methyl-1-butene	C_5H_{10}	3,125.10
35	2-Methyl-2-butene	C_5H_{10}	3,108.62
36	1,2-pentadiene	C_5H_8	3,078.15
37	Cis-1,3-Pentadiene	C_5H_8	2,989.26
38	Trans-1,3-Pentadiene	C_5H_8	2,982.77
39	1,4-pentadiene	C_5H_8	3,015.72

(continued)

(continued)

No.	Compound	Formula	ΔH_c° (net)
40	2,3-pentadiene	C_5H_8	2,351.51
41	3-methyl-1,2-butadiene	C_5H_8	2,626.42
42	2-Methyl-1,3-Butadiene (isoprene)	C_5H_8	3,010.97
43	Cyclopentene	C_5H_8	2,939.51
44	1-pentyne	C_5H_8	3,051.06
45	2-pentyne	C_5H_8	3,029.83
46	3-methyl-1-butyne	C_5H_8	3,045.83
47	Cyclopentadiene	C_5H_6	2,795.30
48	N-Hexane	C_6H_{14}	3,855.14
49	2-methylpentane	C_6H_{14}	3,849.13
50	3-methylpentane	C_6H_{14}	3,851.54
51	2,2-Dimethylbutane (neohexane)	C_6H_{14}	3,841.11
52	2,3-dimethylbutane	C_6H_{14}	3,847.73
53	Ethylcyclobutane	C_6H_{12}	3,788.66
54	Methylcyclopentane	C_6H_{12}	3,673.95
55	Cyclohexane	C_6H_{12}	3,655.94
56	1-hexene	C_6H_{12}	3,739.92
57	Cis-2-Hexene	C_6H_{12}	3,728.37
58	Trans-2-Hexene	C_6H_{12}	3,726.61
59	Cis-3-Hexene	C_6H_{12}	3,733.27
60	Trans-3-Hexene	C_6H_{12}	3,726.22
61	2-Methyl-1-pentene	C_6H_{12}	3,722.30
62	3-Methyl-1-pentene	C_6H_{12}	3,734.05
63	4-Methyl-1-pentene	C_6H_{12}	3,732.29
64	2-Methyl-2-pentene	C_6H_{12}	3,713.49
65	Cis-3-Methyl-2-Pentene	C_6H_{12}	3,717.80
66	Trans-3-Methyl-2-pentene	C_6H_{12}	3,717.80
67	Cis-4-Methyl-2-pentene	C_6H_{12}	3,725.24
68	Trans-4-Methyl-2-pentene	C_6H_{12}	3,720.54
69	2-ethyl-1-butene	C_6H_{12}	3,725.24
70	2,3-dimethyl-1-butene	C_6H_{12}	3,718.19
71	3,3-dimethyl-1-butene	C_6H_{12}	3,724.85
72	2,3-dimethyl-2-butene	C_6H_{12}	3,710.75
73	1,2-hexadiene	C_6H_{10}	3,548.56
74	1,5-hexadiene	C_6H_{10}	3,655.00
75	2,3-hexadiene	C_6H_{10}	3,130.86
76	3-methyl-1,2-pentadiene	C_6H_{10}	3,678.69
77	1-Methyl-cyclopentene	C_6H_{10}	3,532.89
78	Cyclohexene	C_6H_{10}	3,532.13
79	1-hexyne	C_6H_{10}	3,660.92
80	Benzene	C_6H_6	3,135.50
81	N-heptane	C_7H_{16}	4,465.06
82	2-methylhexane	C_7H_{16}	4,459.70
83	3-methylhexane	C_7H_{16}	4,462.73

(continued)

No.	Compound	Formula	ΔH_c° (net)
84	3-ethylpentane	C_7H_{16}	4,464.60
85	2,2-dimethylpentane	C_7H_{16}	4,450.85
86	2,3-dimethylpentane	C_7H_{16}	4,460.87
87	2,4-dimethylpentane	C_7H_{16}	4,454.58
88	3,3-dimethylpentane	C_7H_{16}	4,456.44
89	2,2,3-trimethylbutane (triptane)	C_7H_{16}	4,452.71
90	Ethylcyclopentane	C_7H_{14}	4,284.14
91	1,1-dimethylcyclopentane	C_7H_{14}	4,275.69
92	Cis-1,2-Dimethylcyclopentane	C_7H_{14}	4,282.31
93	Trans-1,2-Dimethylcyclopentane	C_7H_{14}	4,276.37
94	Cis-1,3-Dimethylcyclopentane	C_7H_{14}	4,277.52
95	Trans-1,3-Dimethylcyclopentane	C_7H_{14}	4,279.57
96	Methylcyclohexane	C_7H_{14}	4,257.65
97	Cycloheptane	C_7H_{14}	4,289.85
98	1-heptene	C_7H_{14}	4,385.54
99	Cis-2-Heptene	C_7H_{14}	4,378.69
100	Trans-2-Heptene	C_7H_{14}	4,374.58
101	Cis-3-Heptene	C_7H_{14}	4,378.69
102	Trans-3-Heptene	C_7H_{14}	4,374.58
103	2-Methyl-1-hexene	C_7H_{14}	4,370.47
104	3-Methyl-1-hexene	C_7H_{14}	4,380.98
105	4-Methyl-1-hexene	C_7H_{14}	4,380.98
106	5-Methyl-1-hexene	C_7H_{14}	4,378.24
107	2-Methyl-2-hexene	C_7H_{14}	4,364.53
108	Cis-3-Methyl-2-hexene	C_7H_{14}	4,367.27
109	Trans-3-Methyl-2-hexene	C_7H_{14}	4,367.27
110	Cis-4-Methyl-2-hexene	C_7H_{14}	4,374.35
111	Trans-4-Methyl-2-hexene	C_7H_{14}	4,370.01
112	Cis-5-Methyl-2-Hexene	C_7H_{14}	4,371.61
113	Trans-5-Methyl-2-hexene	C_7H_{14}	4,367.50
114	Cis-2-methyl-3-hexene	C_7H_{14}	4,371.61
115	Trans-2-Methyl-3-hexene	C_7H_{14}	4,367.50
116	Cis-3-Methyl-3-hexene	C_7H_{14}	4,367.27
117	Trans-3-Methyl-3-hexene	C_7H_{14}	4,367.27
118	2-ethyl-1-pentene	C_7H_{14}	4,372.98
119	3-ethyl-1-pentene	C_7H_{14}	4,383.49
120	3-ethyl-2-pentene	C_7H_{14}	4,369.78
121	2,3-Dimethyl-1-pentene	C_7H_{14}	4,366.13
122	2,4-Dimethyl-1-pentene	C_7H_{14}	4,363.16
123	3,3-Dimethyl-1-pentene	C_7H_{14}	4,371.84
124	3,4-Dimethyl-1-pentene	C_7H_{14}	4,374.12
125	4,4-Dimethyl-1-pentene	C_7H_{14}	4,367.04
126	2,3-Dimethyl-2-pentene	C_7H_{14}	4,359.51
127	2,4-Dimethyl-2-pentene	C_7H_{14}	4,357.45

(continued)

(continued)

No.	Compound	Formula	ΔH_c° (net)
128	Cis-3,4-dimethyl-2-pentene	C_7H_{14}	4,360.19
129	Trans-3,4-dimethyl-2-Pentene	C_7H_{14}	4,360.19
130	Cis-4,4-dimethyl-2-pentene	C_7H_{14}	4,360.42
131	Trans-4,4-dimethyl-2-pentene	C_7H_{14}	4,366.13
132	3-Methyl-2-ethyl-1-butene	C_7H_{14}	4,366.13
133	2,3,3-Trimethyl-1-butene	C_7H_{14}	4,360.42
134	2-methyl-1,5-hexadiene	C_7H_{12}	3,639.68
135	2-methyl-2,4-hexadiene	C_7H_{12}	3,500.77
136	1-ethylcyclopentene	C_7H_{12}	4,147.46
137	3-ethylcyclopentene	C_7H_{12}	4,156.41
138	1-methylcyclohexene	C_7H_{12}	4,124.42
139	1-heptyne	C_7H_{12}	4,270.27
140	Toluene	C_7H_8	3,734.06
141	N-Octane	C_8H_{18}	5,074.32
142	2-methylheptane	C_8H_{18}	5,069.54
143	3-methylheptane	C_8H_{18}	5,072.19
144	4-methylheptane	C_8H_{18}	5,072.99
145	3-ethylhexane	C_8H_{18}	5,074.05
146	2,2-dimethylhexane	C_8H_{18}	5,062.63
147	2,3-dimethylhexane	C_8H_{18}	5,071.93
148	2,4-dimethylhexane	C_8H_{18}	5,967.60
149	2,5-Dimethylhexane (diisobutyl)	C_8H_{18}	5,064.22
150	3,3-dimethylhexane	C_8H_{18}	5,067.15
151	3,4-dimethylhexane	C_8H_{18}	5,072.99
152	2-methyl-3-ethylpentane	C_8H_{18}	5,073.26
153	3-methyl-3-ethylpentane	C_8H_{18}	5,071.66
154	2,2,3-trimethylpentane	C_8H_{18}	5,067.68
155	2,2,4-trimethylpentane	C_8H_{18}	5,065.29
156	2,3,3-trimethylpentane	C_8H_{18}	5,068.74
157	2,3,4-trimethylpentane	C_8H_{18}	5,069.54
158	2,2,3,3,-tetramethylbutane	C_8H_{18}	5,963.61
159	N-Propylcyclopentane	C_8H_{16}	4,893.76
160	Isopropylcyclopentane	C_8H_{16}	4,925.86
161	1-methyl-1-ethylcyclopentane	C_8H_{16}	4,923.25
162	Cis-1-Methyl-2-Ethylcyclopentane	C_8H_{16}	4,925.86
163	Trans-1-Methyl-2-Ethylcyclopentane	C_8H_{16}	4,925.86
164	Cis-1-Methyl-3-ethylcyclopentane	C_8H_{16}	4,925.86
165	Trans-1-Methyl-3-ethylcyclopentane	C_8H_{16}	4,925.86
166	1,1,2-trimethylcyclopentane	C_8H_{16}	4,925.86
167	1,1,3-trimethylcyclopentane	C_8H_{16}	4,914.38
168	1,c-2,c-3-Trimethylcyclopentane	C_8H_{16}	4,925.86
169	1,c-2,t-3-Trimethylcyclopentane	C_8H_{16}	4,917.25
170	1,t-2,c-3-Trimethylcyclopentane	C_8H_{16}	4,917.25
171	1,c-2,c-4-Trimethylcyclopentane	C_8H_{16}	4,917.25

(continued)

No.	Compound	Formula	ΔH_c°(net)
172	1,c-2,t-4-Trimethylcyclopentane	C_8H_{16}	4,917.25
173	Ethylcyclohexane	C_8H_{16}	4,870.53
174	1,1-dimethylcyclohexane	C_8H_{16}	4,864.00
175	Cis-1,2-dimethylcyclohexane	C_8H_{16}	4,871.05
176	Trans-1,2-dimethylcyclohexane	C_8H_{16}	4,864.53
177	Cis-1,3-dimethylcyclohexane	C_8H_{16}	4,859.83
178	Trans-1,3-dimethylcyclohexane	C_8H_{16}	4,867.14
179	Cis-1,4-dimethylcyclohexane	C_8H_{16}	4,867.14
180	Trans-1,4-dimethylcyclohexane	C_8H_{16}	4,860.35
181	1,t-2,c-4-trimethylcyclopentane	C_8H_{16}	4,917.25
182	Cyclooctane	C_8H_{16}	4,913.33
183	1-octene	C_8H_{16}	4,959.01
184	Cis-2-octene	C_8H_{16}	4,987.98
185	Trans-2-octene	C_8H_{16}	4,949.87
186	Cis-3-octene	C_8H_{16}	4,988.76
187	Trans-3-octene	C_8H_{16}	4,949.87
188	Cis-4-octene	C_8H_{16}	4,988.76
189	Trans-4-octene	C_8H_{16}	4,949.87
190	2-Methyl-1-heptene	C_8H_{16}	4,985.63
191	3-methyl-1-heptene	C_8H_{16}	4,992.16
192	4-methyl-1-heptene	C_8H_{16}	6,208.68
193	Trans-6-methyl-2-heptene	C_8H_{16}	4,973.89
194	Trans-3-methyl-3-heptene	C_8H_{16}	5,417.15
195	2-ethyl-1-hexene	C_8H_{16}	4,949.87
196	3-ethyl-1-hexene	C_8H_{16}	4,990.33
197	4-ethyl-1-hexene	C_8H_{16}	4,991.11
198	2,3-Dimethyl-1-hexene	C_8H_{16}	5,690.33
199	2,3-Dimethyl-2-hexene	C_8H_{16}	4,959.79
200	Cis-2,2-Dimethyl-3-hexene	C_8H_{16}	4,972.58
201	2,3,3-Trimethyl-1-pentene	C_8H_{16}	4,964.75
202	2,4,4-Trimethyl-1-pentene	C_8H_{16}	4,937.08
203	2,4,4-Trimethyl-2-pentene	C_8H_{16}	4,940.48
204	2,6-octadiene	C_8H_{14}	4,860.44
205	1-n-Propylcyclopentene	C_8H_{14}	4,757.40
206	1-ethylcyclohexene	C_8H_{14}	4,734.58
207	1-octyne	C_8H_{14}	4,880.18
208	Ethylbenzene	C_8H_{10}	4,345.11
209	o-Xylene	C_8H_{10}	4,333.01
210	m-Xylene	C_8H_{10}	4,332.02
211	p-Xylene	C_8H_{10}	4,333.01
212	Styrene	C_8H_8	4,219.56
213	Phenylacetylene	C_8H_6	4,201.32
214	N-nonane	C_9H_{20}	5,684.73
215	2-methyloctane	C_9H_{20}	5,679.36

(continued)

(continued)

No.	Compound	Formula	ΔH_c°(net)
216	3-methyloctane	C_9H_{20}	5,681.45
217	4-methyloctane	C_9H_{20}	5,680.55
218	3-ethylheptane	C_9H_{20}	5,683.84
219	2,2-dimethylheptane	C_9H_{20}	5,671.60
220	2,6-dimethylheptane	C_9H_{20}	5,673.99
221	2,2,3-trimethylhexane	C_9H_{20}	5,677.27
222	2,2,4-trimethylhexane	C_9H_{20}	5,676.97
223	2,2,5-trimethylhexane	C_9H_{20}	5,666.23
224	2,3,3-trimethylhexane	C_9H_{20}	5,678.76
225	2,3,5-trimethylhexane	C_9H_{20}	5,675.78
226	2,4,4-trimethylhexane	C_9H_{20}	5,675.78
227	3,3,4-trimethylhexane	C_9H_{20}	5,682.34
228	3,3-diethylpentane	C_9H_{20}	5,684.43
229	2,2-Dimethyl-3-ethylpentane	C_9H_{20}	5,687.12
230	2,4-Dimethyl-3-ethylpentane	C_9H_{20}	5,690.10
231	2,2,3,3-tetramethylpentane	C_9H_{20}	5,681.75
232	2,2,3,4-tetramethylpentane	C_9H_{20}	5,684.13
233	2,2,4,4-tetramethylpentane	C_9H_{20}	5,679.96
234	2,3,3,4-tetramethylpentane	C_9H_{20}	5,682.05
235	N-Butylcyclopentane	C_9H_{18}	5,504.17
236	Isobutylcyclopentane	C_9H_{18}	5,529.43
237	1-Methyl-1-n-propylcyclopentane	C_9H_{18}	5,537.94
238	1,1-diethylcyclopentane	C_9H_{18}	5,537.94
239	Cis-1,2-diethylcyclopentane	C_9H_{18}	5,540.58
240	1,1-dimethyl-2-ethylcyclopentane	C_9H_{18}	5,529.43
241	N-propylcyclohexane	C_9H_{18}	5,480.10
242	Isopropylcyclohexane	C_9H_{18}	5,512.98
243	Cyclononane	C_9H_{18}	5,535.59
244	Ethylcycloheptane	C_9H_{18}	5,549.98
245	1-nonene	C_9H_{18}	5,568.48
246	2,6-dimethyl-1,5-heptadiene	C_9H_{16}	5,021.82
247	1-nonyne	C_9H_{16}	5,490.22
248	N-propylbenzene	C_9H_{12}	4,954.12
249	Isopropylbenzene (cumene)	C_9H_{12}	4,951.32
250	O-Ethyltoluene	C_9H_{12}	4,946.01
251	M-Ethyltoluene	C_9H_{12}	4,943.77
252	P-Ethyltoluene	C_9H_{12}	4,942.66
253	1,2,3-Trimethylbenzene (hemimellitene)	C_9H_{12}	4,933.99
254	1,2,4-Trimethylbenzene (pseudocumene)	C_9H_{12}	4,930.63
255	1,3,5-Trimethylbenzene (mesitylene)	C_9H_{12}	4,928.96
256	Cis-1-propenyl benzene	C_9H_{10}	4,872.64
257	Trans-1-propenyl benzene	C_9H_{10}	4,868.52
258	2-Propenyl benzene	C_9H_{10}	4,818.21
259	Indan (2,3-dihydroindene)	C_9H_{10}	4,790.73

(continued)

No.	Compound	Formula	ΔH_c°(net)
260	1-methyl-2-ethynyl benzene	C_9H_8	4,869.62
261	1-methyl-3-ethenyl benzene	C_9H_8	4,866.59
262	1-methyl-4-ethenyl benzene	C_9H_8	4,865.77
263	Indene	C_9H_8	4,619.41
264	N-decane	$C_{10}H_{22}$	6,294.21
265	2-methylnonane	$C_{10}H_{22}$	6,288.92
266	3-methylnonane	$C_{10}H_{22}$	6,291.23
267	4-methylnonane	$C_{10}H_{22}$	6,291.23
268	5-methylnonane	$C_{10}H_{22}$	6,291.23
269	2,7-dimethyloctane	$C_{10}H_{22}$	6,283.62
270	3,3,4-trimethylheptane	$C_{10}H_{22}$	6,291.23
271	3,3,5-trimethylheptane	$C_{10}H_{22}$	6,290.57
272	2,2,3,3-tetramethylhexane	$C_{10}H_{22}$	6,291.56
273	2,2,5,5-tetramethylhexane	$C_{10}H_{22}$	6,267.73
274	2,4-dimethyl-3-isopropylpentane	$C_{10}H_{22}$	6,311.42
275	N-Penthylcyclopentane	$C_{10}H_{20}$	6,113.94
276	N-Butylcyclohexane	$C_{10}H_{20}$	6,090.77
277	Iso-butylcyclohexane	$C_{10}H_{20}$	6,127.64
278	Sec-butylcyclohexane	$C_{10}H_{20}$	6,127.64
279	Tert-butylcyclohexane	$C_{10}H_{20}$	6,116.22
280	1-methyl-4-isopropylcyclohexane	$C_{10}H_{20}$	6,118.83
281	1-decene	$C_{10}H_{20}$	6,178.21
282	Cis-decahydronaphthalene (cis-decalin)	$C_{10}H_{18}$	5,892.12
283	Trans-decahydronaphthalene (trans-decalin)	$C_{10}H_{18}$	5,880.86
284	3,7-dimethyl-1,6-octadiene	$C_{10}H_{18}$	6,079.59
285	1-decyne	$C_{10}H_{18}$	6,099.53
286	N-butylbenzene	$C_{10}H_{14}$	5,564.58
287	Isobutylbenzene	$C_{10}H_{14}$	5,558.02
288	Sec-Butylbenzene	$C_{10}H_{14}$	5,561.45
289	Tert-Butylbenzene	$C_{10}H_{14}$	5,557.08
290	1-Methyl-2-n-propylbenzene	$C_{10}H_{14}$	5,555.52
291	1-Methyl-3-n-propylbenzene	$C_{10}H_{14}$	5,551.78
292	1-Methyl-4-n-Propylbenzene	$C_{10}H_{14}$	5,552.71
293	o-Cymene	$C_{10}H_{14}$	5,554.59
294	m-Cymene	$C_{10}H_{14}$	5,549.28
295	p-Cymene	$C_{10}H_{14}$	5,549.90
296	o-Diethylbenzene	$C_{10}H_{14}$	5,559.27
297	m-Diethylbenzene	$C_{10}H_{14}$	5,554.27
298	p-Diethylbenzene	$C_{10}H_{14}$	5,555.21
299	1,2-Dimethyl-3-ethylbenzene	$C_{10}H_{14}$	5,547.41
300	1,2-dimethyl-4-ethylbenzene	$C_{10}H_{14}$	5,541.79
301	1,3-dimethyl-2-ethylbenzene	$C_{10}H_{14}$	5,547.72
302	1,3-Dimethyl-4-ethylbenzene	$C_{10}H_{14}$	5,592.36

(continued)

(continued)

No.	Compound	Formula	ΔH_c° (net)
303	1,3-Dimethyl-5-ethylbenzene	$C_{10}H_{14}$	5,592.36
304	1,4-Dimethyl-2-ethylbenzene	$C_{10}H_{14}$	5,592.36
305	1,2,3,4-tetramethylbenzene	$C_{10}H_{14}$	5,537.73
306	1,2,3,5-tetramethylbenzene	$C_{10}H_{14}$	5,531.48
307	1,2,4,5-tetramethylbenzene	$C_{10}H_{14}$	5,508.07
308	Dicyclopentadiene	$C_{10}H_{12}$	5,551.25
309	1-Methyl-4-(trans-1-n-Propenyl)Benzene	$C_{10}H_{12}$	5,468.22
310	1-ethyl-2-ethenyl benzene	$C_{10}H_{12}$	5,479.91
311	1-ethyl-3-ethenyl benzene	$C_{10}H_{12}$	5,479.91
312	1-ethyl-4-ethenyl benzene	$C_{10}H_{12}$	5,479.91
313	2-phenyl-1-butene	$C_{10}H_{12}$	5,478.99
314	1,2,3,4-tetrahydronaphthalene	$C_{10}H_{12}$	5,357.52
315	1-methyl-2,3-dihydroindene	$C_{10}H_{12}$	5,399.65
316	2-methyl-2,3-dihydroindene	$C_{10}H_{12}$	5,399.65
317	4-methyl-2,3-dihydroindene	$C_{10}H_{12}$	5,392.89
318	5-methyl-2,3-dihydroindene	$C_{10}H_{12}$	5,393.19
319	1-methylindene	$C_{10}H_{10}$	4,761.57
320	Naphthalene	$C_{10}H_8$	4,980.75
321	N-undecane	$C_{11}H_{24}$	6,903.60
322	N-hexylcyclopentane	$C_{11}H_{22}$	6,724.03
323	N-pentylcyclohexane	$C_{11}H_{22}$	6,700.70
324	1-undecene	$C_{11}H_{22}$	6,788.27
325	N-pentylbenzene	$C_{11}H_{16}$	6,174.17
326	1-methyl-[1,2,3,4-tetrahydronaphthalene]	$C_{11}H_{14}$	6,014.88
327	1-methylnaphthalene	$C_{11}H_{10}$	5,594.10
328	2-methylnaphthalene	$C_{11}H_{10}$	5,582.85
329	N-dodecane	$C_{12}H_{26}$	7,513.74
330	N-heptylcyclopentane	$C_{12}H_{24}$	7,333.42
331	N-hexylcyclohexane	$C_{12}H_{24}$	7,310.32
332	1-dodecene	$C_{12}H_{24}$	7,397.63
333	Bicyclohexyl	$C_{12}H_{22}$	7,053.20
334	N-hexylbenzene	$C_{12}H_{18}$	6,783.73
335	Cyclohexylbenzene	$C_{12}H_{16}$	6,575.20
336	1-ethyl-[1,2,3,4-tetrahydronaphthalene]	$C_{12}H_{16}$	6,629.62
337	2,2-dimethyl-[1,2,3,4-tetrahydronaphthalene]	$C_{12}H_{16}$	6,405.22
338	2,6-dimethyl-[1,2,3,4-tetrahydronaphthalene]	$C_{12}H_{16}$	6,613.97
339	6,7-dimethyl-[1,2,3,4-tetrahydronaphthalene]	$C_{12}H_{16}$	6,610.24
340	1-ethylnaphthalene	$C_{12}H_{12}$	6,271.76
341	2-ethylnaphthalene	$C_{12}H_{12}$	6,499.97
342	1,2-dimethylnaphthalene	$C_{12}H_{12}$	6,487.98
343	1,4-dimethylnaphtalene	$C_{12}H_{12}$	6,487.98
344	Biphenyl	$C_{12}H_{10}$	6,031.77
345	Acenaphthene	$C_{12}H_{10}$	6,001.64
346	Acenaphthalene	$C_{12}H_8$	5,848.69

(continued)

No.	Compound	Formula	ΔH_c°(net)
347	N-tridecane	$C_{13}H_{28}$	8,122.73
348	N-octylcyclopentane	$C_{13}H_{26}$	7,943.41
349	N-heptylcyclohexane	$C_{13}H_{26}$	7,920.08
350	1-tridecene	$C_{13}H_{26}$	8,007.45
351	N-heptylbenzene	$C_{13}H_{20}$	7,393.63
352	1-n-Propyl-[1,2,3,4-Tetrahydronaphthalene]	$C_{13}H_{18}$	6,817.59
353	6-n-Propyl-[1,2,3,4-Tetrahydronaphthalene]	$C_{13}H_{18}$	5,033.83
354	1-n-Propylnaphthalene	$C_{13}H_{14}$	7,114.56
355	2-n-Propylnaphthalene	$C_{13}H_{14}$	7,114.56
356	1-methyl-2-phenylbenzene	$C_{13}H_{12}$	6,713.59
357	1-methyl-3-phenylbenzene	$C_{13}H_{12}$	6,713.59
358	1-methyl-4-phenylbenzene	$C_{13}H_{12}$	6,713.59
359	Diphenylmethane	$C_{13}H_{12}$	6,663.89
360	Fluorene	$C_{13}H_{10}$	6,424.98
361	N-tetradecane	$C_{14}H_{30}$	8,733.04
362	N-nonylcyclopentane	$C_{14}H_{28}$	8,553.66
363	N-octylcyclohexane	$C_{14}H_{28}$	8,530.36
364	1-tetradecene	$C_{14}H_{28}$	8,617.15
365	N-octylbenzene	$C_{14}H_{22}$	8,003.27
366	1-ethyl-4-phenylbenzene	$C_{14}H_{14}$	7,328.17
367	1-methyl-4(4-methylphenyl)-benzene	$C_{14}H_{14}$	6,601.97
368	1,1-diphenylethane	$C_{14}H_{14}$	7,249.74
369	1,2-diphenylethane	$C_{14}H_{14}$	7,243.38
370	1-n-Butylnaphthalene	$C_{14}H_{16}$	7,729.15
371	2-n-Butylnaphthalene	$C_{14}H_{16}$	7,729.15
372	Cis-1,2-diphenylethene	$C_{14}H_{12}$	7,197.04
373	Trans-1,2-diphenylethene	$C_{14}H_{12}$	7,201.24
374	Diphenylacetylene	$C_{14}H_{10}$	7,148.31
375	Anthracene	$C_{14}H_{10}$	6,847.34
376	Phenanthrene	$C_{14}H_{10}$	6,834.49
377	N-pentadecane	$C_{15}H_{32}$	9,342.73
378	N-decylcyclopentane	$C_{15}H_{30}$	9,162.86
379	N-nonylcyclohexane	$C_{15}H_{30}$	9,139.86
380	1-pentadecene	$C_{15}H_{30}$	9,226.48
381	N-nonylbenzene	$C_{15}H_{24}$	8,613.19
382	1-n-Pentylnaphthalene	$C_{15}H_{18}$	8,343.90
383	1,2-diphenylpropane	$C_{15}H_{16}$	7,954.83
384	N-hexadecane	$C_{16}H_{34}$	9,951.46
385	N-undecylcyclopentane	$C_{16}H_{32}$	9,772.81
386	N-decylcyclohexane	$C_{16}H_{32}$	9,749.85
387	1-hexadecene	$C_{16}H_{32}$	9,836.50
388	N-decylbenzene	$C_{16}H_{26}$	9,222.37
389	1-n-Hexyl-[1,2,3,4-Tetrahydronaphthalene]	$C_{16}H_{24}$	9,088.17
390	1-n-Hexylnaphthalene	$C_{16}H_{20}$	8,958.48

(continued)

(continued)

No.	Compound	Formula	ΔH_c° (net)
391	2-n-Hexylnaphthalene	$C_{16}H_{20}$	8,958.48
392	Pyrene	$C_{16}H_{10}$	7,620.08
393	Fluoranthene	$C_{16}H_{10}$	7,694.88
394	N-heptadecane	$C_{17}H_{36}$	10,561.89
395	N-dodecylcyclopentane	$C_{17}H_{34}$	10,383.20
396	N-undecylcyclohexane	$C_{17}H_{34}$	10,359.90
397	1-heptadecene	$C_{17}H_{34}$	10,446.43
398	N-undecylbenzene	$C_{17}H_{28}$	9,832.17
399	1-n-Heptyl-[1,2,3,4-Tetrahydronaphthalene]	$C_{17}H_{26}$	9,702.77
400	1-n-Heptylnaphthalene	$C_{17}H_{22}$	9,573.07
401	N-octadecane	$C_{18}H_{38}$	11,171.60
402	N-tridecylcyclopentane	$C_{18}H_{36}$	10,992.49
403	N-dodecylcyclohexane	$C_{18}H_{36}$	10,969.00
404	1-octadecene	$C_{18}H_{36}$	11,055.92
405	N-dodecylbenzene	$C_{18}H_{30}$	10,442.34
406	1-n-Octyl-[1,2,3,4-Tetrahydronaphthalene]	$C_{18}H_{28}$	10,317.52
407	1-n-Octylnaphthalene	$C_{18}H_{24}$	10,187.66
408	1,2-Diphenylbenzene (o-terphenyl)	$C_{18}H_{14}$	9,052.81
409	1,3-Diphenylbenzene (m-terphenyl)	$C_{18}H_{14}$	9,052.81
410	1,4-Diphenylbenzene (p-terphenyl)	$C_{18}H_{14}$	9,052.81
411	Chrysene	$C_{18}H_{12}$	8,679.24
412	N-nonadecane	$C_{19}H_{40}$	11,781.22
413	N-tetradecylcyclopentane	$C_{19}H_{38}$	11,602.71
414	N-tridecylcyclohexane	$C_{19}H_{38}$	11,579.15
415	1-nonadecene	$C_{19}H_{38}$	11,665.94
416	N-tridecylbenzene	$C_{19}H_{32}$	11,051.55
417	1-n-Nonyl-[1,2,3,4-Tetrahydronaphthalene]	$C_{19}H_{30}$	10,932.00
418	1-n-Nonylnaphthalene	$C_{19}H_{26}$	10,500.15
419	2-n-Nonylnapthalene	$C_{19}H_{26}$	10,801.94
420	N-eicosane	$C_{20}H_{42}$	12,391.06
421	N-pentadecylcyclopentane	$C_{20}H_{40}$	12,212.21
422	N-tetradecylcyclohexane	$C_{20}H_{40}$	12,189.37
423	1-eicosene	$C_{20}H_{40}$	12,275.51
424	N-tetradecylbenzene	$C_{20}H_{34}$	11,661.54
425	1-n-Decyl-[1,2,3,4-Tetrahydronaphthalene]	$C_{20}H_{32}$	11,546.57
426	1-n-Decylnaphthalene	$C_{20}H_{28}$	11,100.43
427	N-heneicosane	$C_{21}H_{44}$	13,106.37
428	N-hexadecylcyclopentane	$C_{21}H_{42}$	12,821.83
429	N-pentadecylcyclohexane	$C_{21}H_{42}$	12,798.54
430	N-pentadecylbenzene	$C_{21}H_{36}$	12,271.02
431	N-docosane	$C_{22}H_{46}$	13,721.32
432	N-heptadecylcyclopentane	$C_{22}H_{44}$	13,539.49
433	N-hexadecylcyclohexane	$C_{22}H_{44}$	13,408.86
434	N-hexadecylbenzene	$C_{22}H_{38}$	12,880.67

(continued)

No.	Compound	Formula	ΔH_c°(net)
435	N-tricosane	$C_{23}H_{48}$	14,335.37
436	N-octadecylcyclopentane	$C_{23}H_{46}$	14,154.31
437	N-heptadecylcyclohexane	$C_{23}H_{46}$	14,126.55
438	N-tetracosane	$C_{24}H_{50}$	14,950.20
439	N-nonadecylcyclopentane	$C_{24}H_{48}$	14,768.63
440	N-octedecylcyclohexane	$C_{24}H_{48}$	14,741.22
441	N-pentacosane	$C_{25}H_{52}$	15,564.63
442	N-eicosylcyclopentane	$C_{25}H_{50}$	15,383.32
443	N-nonadecylcyclohexane	$C_{25}H_{50}$	15,355.59
444	N-hexacosane	$C_{26}H_{54}$	16,179.94
445	N-eicocylcyclohexane	$C_{26}H_{52}$	15,970.80
446	N-heptacosane	$C_{27}H_{56}$	16,794.54
447	N-octacosane	$C_{28}H_{58}$	17,408.82
448	N-nonacosane	$C_{29}H_{60}$	18,023.28
449	N-triacontane	$C_{30}H_{62}$	18,637.92
450	Formic acid	CH_2O_2	211.45
451	Acetic acid	$C_2H_4O_2$	786.38
452	Propionic acid	$C_3H_6O_2$	1,395.02
453	N-butyric acid	$C_4H_8O_2$	2,006.40
454	2-methylpropionic acid	$C_4H_8O_2$	2,000.25
455	N-pentanoic acid	$C_5H_{10}O_2$	2,616.42
456	2-methylbutyric acid	$C_5H_{10}O_2$	2,680.09
457	3-methylbutyric acid	$C_5H_{10}O_2$	2,615.24
458	N-hexanoic acid	$C_6H_{12}O_2$	3,228.75
459	Methanol	CH_4O	638.08
460	Ethanol	C_2H_6O	1,235.54
461	N-propanol	C_3H_8O	1,844.00
462	Isopropanol	C_3H_8O	1,830.16
463	N-butanol	$C_4H_{10}O$	2,456.05
464	Isobutanol	$C_4H_{10}O$	2,448.99
465	Sec-butanol	$C_4H_{10}O$	2,440.54
466	Tert-butanol	$C_4H_{10}O$	2,423.82
467	1-pentanol	$C_5H_{12}O$	3,060.59
468	2-pentanol	$C_5H_{12}O$	3,051.56
469	2-methyl-1-butanol	$C_5H_{12}O$	3,062.02
470	2-methyl-2-butanol	$C_5H_{12}O$	3,039.26
471	3-methyl-2-butanol	$C_5H_{12}O$	3,052.18
472	2,2-dimethy-1-propanol	$C_5H_{12}O$	3,099.54
473	4-methyl-2-pentanol	$C_6H_{14}O$	3,661.55
474	Phenol	C_6H_6O	2,921.44
475	o-Cresol	C_7H_8O	3,517.45
476	m-Cresol	C_7H_8O	3,528.01
477	p-Cresol	C_7H_8O	3,522.73
478	Formaldehyde	CH_2O	519.54

(continued)

(continued)

No.	Compound	Formula	ΔH_c° (net)
479	Acetaldehyde	C_2H_4O	1,104.52
480	N-propionaldehyde	C_3H_6O	1,685.70
481	Acrolein	C_3H_4O	1,553.53
482	N-butyraldehyde	C_4H_8O	2,303.57
483	Trans-Crotonaldehyde	C_4H_6O	2,155.57
484	Methacrolein	C_4H_6O	2,154.76
485	Methylamine	CH_5N	975.17
486	Ethylamine	C_2H_7N	1,587.31
487	N-propylamine	C_3H_9N	2,164.78
488	Isopropylamine	C_3H_9N	2,156.67
489	N-butylamine	$C_4H_{11}N$	2,776.42
490	Isobutylamine	$C_4H_{11}N$	2,771.82
491	Sec-bytylamine	$C_4H_{11}N$	2,766.55
492	Tert-butylamine	$C_4H_{11}N$	2,753.79
493	Urea	CH_4N_2O	544.27
494	Acetonitrile	C_2H_3N	1,190.38
495	Morpholine	C_4H_9NO	2,460.06
496	Pyridine	C_5H_5N	2,672.04
497	Aniline	C_6H_7N	3,238.69
498	Indole	C_8H_7N	4,081.10
499	Quinoline	C_9H_7N	4,544.25
500	Dibenzopyrrole	$C_{12}H_9N$	5,935.86
501	Acridine	$C_{13}H_9N$	6,403.89
502	Methyl formate	$C_2H_4O_2$	920.89
503	Methyl acetate	$C_3H_6O_2$	1,461.02
504	Ethyl formate	$C_3H_6O_2$	1,507.02
505	Ethyl acetate	$C_4H_8O_2$	2,062.96
506	N-propyl formate	$C_4H_8O_2$	2,041.04
507	Vinyl acetate	$C_4H_6O_2$	1,948.99
508	Methyl n-butyrate	$C_5H_{10}O_2$	2,692.68
509	N-propyl acetate	$C_5H_{10}O_2$	2,713.11
510	Isopropyl acetate	$C_5H_{10}O_2$	2,658.00
511	N-butyl acetate	$C_6H_{12}O_2$	3,282.79
512	N-petyl acetate	$C_7H_{14}O_2$	3,893.38
513	Dimethy ether	C_2H_6O	1,328.45
514	Methyl ethyl ether	C_3H_8O	1,931.51
515	Diethyl ether	$C_4H_{10}O$	2,503.47
516	Methyl-tert-butyl ether	$C_{10}H_{22}O$	3,099.95
517	Tetrahydrofuran	C_4H_8O	2,325.21
518	Dibenzofuran	$C_{12}H_8O$	5,696.80
519	Carbon monoxide	CO	283.02
520	Carbonyl sulfide	COS	548.27
521	Chlorotrifluoromethane	$CClF_3$	314.41
522	Dichlorodifluoromethane	CCl_2F_2	98.15

(continued)

No.	Compound	Formula	ΔH_c°(net)
523	Trichlorofluoromethane	CCl_3F	104.80
524	Carbon tetrachloride	CCl_4	257.96
525	Carbon tetrafluoride	CF_4	539.76
526	Chlorodifluoromethane	$CHClF_2$	65.77
527	Dichlorofluoromethane	$CHCl_2F$	231.25
528	Chloroform	$CHCl_3$	379.86
529	Trifluoromethane	CHF_3	178.80
530	Dichloromethane	CH_2Cl_2	513.82
531	Methyl chloride	CH_3Cl	675.40
532	Methyl fluoride	CH_3F	521.94
533	Vinyl chloride	C_2H_3Cl	1,158.06
534	1,1,1-trichloroethane	$C_2H_3Cl_3$	974.93
535	1,1,2-trichloroethane	$C_2H_3Cl_3$	963.76
536	1,1,1-trifluoroethane	$C_2H_3F_3$	404.05
537	1,1-dichloroethane	$C_2H_4Cl_2$	1,109.70
538	1,2-dichloroethane	$C_2H_4Cl_2$	1,105.10
539	1,1-difluoroethane	$C_2H_4F_2$	758.48
540	Ethyl chloride	C_2H_5Cl	1,284.88
541	Ethyl fluoride	C_2H_5F	1,110.05
542	1,2-dichloropropane	$C_3H_6Cl_2$	1,704.62
543	Acetone	C_3H_6O	1,659.23
544	Methyl ethyl ketone	C_4H_8O	2,261.81
545	Diethyl ketone	$C_5H_{10}O$	2,880.26
546	Methyl-n-propyl ketone	$C_9H_{18}O$	2,879.86
547	Methyl-n-butyl ketone	$C_{11}H_{22}O$	3,490.16
548	Methyl isobutyl ketone	$C_6H_{12}O$	3,487.36
549	Carbon disulfide	CS_2	1,104.26
550	Methyl mercaptan	CH_4S	1,151.59
551	2,3-dithiabutane	$C_2H_6S_2$	2,043.85
552	Dimethyl sulfide	C_2H_6S	1,744.00
553	Ethyl mercaptan	C_2H_6S	1,735.76
554	N-butanethiol	$C_4H_{10}S$	2,955.29
555	Tert-butanethiol	$C_4H_{10}S$	2,939.56
556	2-butanethiol	$C_4H_{10}S$	3,473.39
557	2-methyl-1-propanethiol	$C_4H_{10}S$	3,472.52
558	3-thiapentene	C_5H_8S	2,960.74
559	2- thiahexane	$C_5H_{12}S$	4,140.55
560	3- thiahexane	$C_5H_{12}S$	4,113.88
561	1-pentanethiol	$C_5H_{12}S$	3,565.84
562	Furfural	$C_5H_4O_2$	2,249.80
563	1,2-propylene glycol	$C_3H_8O_2$	1,647.56
564	Diethylene glycol	$C_4H_{10}O_3$	2,154.87
565	Tetraethylene glycol	$C_8H_{18}O_5$	4,380.00
566	Monoethanolamine	C_2H_7NO	1,363.04

(continued)

(continued)

No.	Compound	Formula	ΔH_c° (net)
567	Diethanolamine	$C_4H_{11}NO_2$	2,410.58
568	Methyl diethanolamine	$C_5H_{13}NO_2$	3,059.91
569	Triethanolamine	$C_6H_{15}NO_3$	3,510.76
570	N,n-dimethylformamide	C_3H_7NO	1,788.65
571	N-methyl-2-pyrrolidone	C_5H_9NO	2,805.19
572	Dimethyl sulfuxide	C_2H_6OS	1,547.43
573	Sulfolane	$C_4H_8O_2S$	2,380.07
574	Formic acid	CH_2O_2	211.45

Appendix D

Experimental values of heat release capacity ($J\,g^{-1}\,K^{-1}$), total heat release ($kJ\,g^{-1}$) and char yield (%)

Name	Elemental composition in repeat unit composition	Total heat release	Char yield	heat release capacity
Polyethylene (PE)	C_2H_4	41.6 [236]	0 [241]	1676 [241]
Polyoxymethylene (POM)	CH_2O	14.0 [236]	0 [241]	169 [241]
Polypropylene (PP)	C_3H_6	41.4 [236]	0 [241]	1571 [241]
Poly(vinyl alcohol) (99%; PVOH)	C_2H_4O	21.6 [236]	3.3 [241]	533 [241]
Poly(ethylene oxide)	C_2H_4O	21.6 [236]	1.7 [241]	652 [241]
Polyisobutylene	C_4H_8	44.4 [236]	0 [241]	1002 [241]
Poly(vinyl chloride)	C_2H_3Cl	11.3 [236]	15.3 [241]	138 [241]
Poly(vinylidene fluoride)	$C_2H_2F_2$	9.7 [236]	7.0 [241]	311 [241]
Polyacrylamide	C_3H_5NO	13.3 [236]	8.3 [241]	104 [241]
Poly(acrylic acid)	$C_3H_4O_2$	12.5 [236]	6.1 [241]	165 [241]
Poly(vinyl acetate) (PVAc)	$C_4H_6O_2$	19.2 [236]	1.2 [241]	313 [241]
Poly(methacrylic acid)	$C_4H_6O_2$	18.4 [236]	0.5 [241]	464 [241]
Polychloroprene	C_4H_5Cl	16.1 [236]	12.9 [241]	188 [241]
Poly(tetrafluoro ethylene) (PTFE)	C_2F_4	3.7 [236]	0 [241]	35 [241]
Poly(methyl methacrylate) (PMMA)	$C_5H_8O_2$	24.3 [236]	0 [241]	514 [241]
Poly(methyl methacrylate) (PMMA)	$C_5H_8O_2$	23.2 [236]	0 [241]	461 [241]
Poly(ethyl acrylate)	$C_5H_8O_2$	22.6 [236]	0.3 [241]	323 [241]
Polymethacrylamide	$C_4H_7NO_2$	18.7 [236]	4.5 [241]	103 [241]
Polystyrene (PS)	C_8H_8	38.8 [236]	0 [241]	927 [241]
Isotactic polystyrene	C_8H_8	39.9 [236]	0 [241]	880 [241]
Poly(2-vinyl pyridene)	C_7H_7N	37.4 [236]	0 [241]	612 [241]
Poly(4-vinyl pyridene)	C_7H_7N	31.7 [236]	0 [241]	568 [241]
Poly(1,4-phenylene sulfide) (PPS)	C_6H_4S	17.1 [236]	41.6 [241]	165 [241]
Poly(n-vinyl pyrrolidone)	C_6H_9NO	25.1 [236]	0 [241]	332 [241]
Polycaprolactam	$C_6H_{11}NO$	28.7 [236]	0 [241]	487 [241]
Polycaprolactone	$C_6H_{10}O_2$	24.4 [236]	0 [241]	526 [241]
Poly(ethyl methacrylate)	$C_6H_{10}O_2$	26.4 [236]	0 [241]	470 [241]
Poly(ethyl methacrylate)	$C_6H_{10}O_2$	26.8 [236]	0 [241]	380 [241]
Poly(α-methyl styrene)	C_9H_{10}	35.5 [236]	0 [241]	730 [241]
Poly(2,6-dimethyl 1,4 phenyleneoxide) (PPO)	C_8H_8O	20.0 [236]	25.5 [241]	409 [241]

(continued)

(continued)

Name	Elemental composition in repeat unit composition	Total heat release	Char yield	heat release capacity
Poly(4-vinyl phenol)	C_8H_8O	27.6 [236]	2.8 [241]	261 [241]
Poly(ethylene maleic anhydride)	$C_6H_6O_3$	12.1 [236]	2.8 [241]	138 [241]
Poly(vinyl butyral)	$C_8H_{14}O_2$	26.9 [236]	0.1 [241]	806 [241]
Poly(2-vinyl naphthalene)	$C_{12}H_{10}$	39.0 [236]	0 [241]	834 [241]
Poly(benzoyl 1,4phenylene)	$C_{13}H_8O$	10.9 [236]	65.2 [241]	41 [241]
Poly(ethylene terephthalate) (PET)	$C_{10}H_8O_4$	15.3 [236]	5.1 [241]	332 [241]
Poly(ether ketone) (PEK)	$C_{13}H_8O_2$	10.8 [236]	59.2 [241]	124 [241]
Polylaurolactam	$C_{12}H_{23}ON$	33.2 [236]	0 [241]	743 [241]
Poly(styrene maleic anhydride)	$C_{12}H_{10}O_3$	23.3 [236]	2.2 [241]	279 [241]
Poly(acrylonitrile butadiene styrene) (ABS)	$C_{15}H_{17}N$	36.6 [236]	0 [241]	669 [241]
Poly(1,4-butanediol terephthalate) (PBT)	$C_{12}H_{12}O_4$	20.3 [236]	1.5 [241]	474 [241]
Poly(hexamethylene adipamide)	$C_{12}H_{12}O_2N_2$	27.4 [236]	0 [241]	615 [241]
Polyazomethine	$C_{15}H_9N_3$	8.7 [236]	77.8 [241]	36 [241]
Poly(1,4-phenylene ether sulfone) (PES)	$C_{12}H_8O_3S$	11.2 [236]	29.3 [241]	115 [241]
Poly(p-phenylene benzobisoxazole) (PBO)	$C_{14}H_6O_2N_2$	5.4 [236]	69.5 [241]	42 [241]
Poly(p-phenylene terephthalamide)	$C_{14}H_{10}O_2N_2$	14.8 [236]	36.1 [241]	302 [241]
Poly(m-phenylene isophthalamide)	$C_{14}H_{10}O_2N_2$	11.7 [236]	48.4 [241]	52 [241]
Poly(ethylene naphthylate) (PEN)	$C_{14}H_{10}O_4$	16.8 [236]	18.2 [241]	309 [241]
Dicyclopentadienyl bisphenol cyanate ester	$C_{17}H_{17}NO$	20.1 [236]	27.1 [241]	493 [241]
Polycarbonate of bisphenol A (PC)	$C_{16}H_{14}O_3$	16.3 [236]	27.1 [241]	359 [241]
Polyphosphazene	$C_{14}H_{14}PNO_3$	21.9 [236]	20 [241]	204 [241]
Poly(dichloroethyl diphenyl ether)	$C_{14}H_8OC_{l2}$	5.2 [236]	57.1 [241]	16 [241]
Cyano-substituted Kevlar	$C_{15}H_9N_3O_2$	9.1 [236]	58.3 [241]	54 [241]
Bisphenol E polycyanurate	$C_{16}H_{12}O_2N_2$	14.7 [236]	41.9 [241]	316 [241]
Bisphenol A polycyanurate	$C_{17}H_{14}O_2N_2$	17.6 [236]	36.3 [241]	283 [241]
Poly(hexamethylene sebacamide)	$C_{16}H_{30}O_2N_2$	35.7 [236]	0 [241]	878 [241]

(continued)

Name	Elemental composition in repeat unit composition	Total heat release	Char yield	heat release capacity
Poly(ether ether ketone) (PEEK)	$C_{19}H_{12}O_3$	12.4 [236]	46.5 [241]	155 [241]
Poly(siloxytetraalkyl biphenylene oxide) (PSA)	$C_{18}H_{18}SiO_2$	15.7 [236]	60.1 [241]	119 [241]
Poly(ether ketoneketone) (PEKK)	$C_{20}H_{12}O_3$	8.7 [236]	60.7 [241]	96 [241]
Tetramethyl bisphenol F polycyanurate	$C_{19}H_{18}O_2N_2$	17.4 [236]	35.4 [241]	280 [241]
Bisphenol C polycarbonate	$C_{15}H_8O_3Cl_2$	3.0 [236]	50.1 [241]	29 [241]
Polybenzimidazole (PBI)	$C_{20}H_{12}N_4$	8.6 [236]	67.5 [241]	36 [241]
Poly(hexamethylene dodecane diamide)	$C_{18}H_{34}N_2O_2$	30.8 [236]	0 [241]	707 [241]
Bisphenol C, polycyanurate	$C_{16}H_8O_2Cl_2$	4.2 [236]	53.3 [241]	24 [241]
Bisphenol A epoxy, catalytic cure phenoxy A	$C_{21}H_{24}O_4$	26.0 [236]	3.9 [241]	657 [241]
Phenolphthalein polycarbonate	$C_{21}H_{12}O_5$	8.0 [236]	49.8 [241]	28 [241]
Poly(amide imide) (PAI)	$C_{22}H_{14}O_3N_2$	7.1 [236]	53.6 [241]	33 [241]
Novolac polycyanurate	$C_{23}H_{15}O_3N_3$	9.9 [236]	51.9 [241]	122 [241]
Polyimide (PI)	$C_{22}H_{10}O_5N_2$	6.6 [236]	51.9 [241]	25 [241]
Hexafluorobisphenol A polycyanurate	$C_{17}H_8O_2N_2F_6$	2.3 [236]	55.2 [241]	32 [241]
Bisphenol C epoxy	$C_{20}H_{18}O_4Cl_2$	10.0 [236]	36 [241]	506 [241]
Bisphenol M polycyanurate	$C_{26}H_{24}O_2N_2$	22.5 [236]	26.4 [241]	239 [241]
Poly(phenyl sulfone)	$C_{24}H_{16}SO_4$	11.3 [236]	38.4 [241]	153 [241]
Bisphenol C polyarylate	$C_{22}H_{12}O_4Cl_2$	7.6 [236]	42.7 [241]	21 [241]
Biphenol phthalonitrile	$C_{28}H_{14}N_4O_2$	3.5 [236]	78.8 [241]	15 [241]
Polysulfone of bisphenol A PSF	$C_{27}H_{22}O_4S$	19.4 [236]	28.1 [241]	345 [241]
LaRC-1A	$C_{28}H_{14}N_2O_6$	6.7 [236]	57.0 [241]	38 [241]
Epoxy Novolac, catalytic cure phenoxy N	$C_{10}H_{11}O$	18.9 [236]	15.9 [241]	246 [241]
Bisphenol A phthalonitrile	$C_{31}H_{20}N_4O_2$	5.9 [236]	73.6 [241]	40 [241]
Technora	$C_{34}H_{24}N_4O_5$	15.3 [236]	41.8 [241]	131 [241]
Bisphenol A6F phthalonitrile	$C_{31}H_{14}N_4O_2F_6$	2.8 [236]	63.8 [241]	9 [241]
Poly(ether imide) (PEI)	$C_{37}H_{24}O_6N_2$	11.8 [236]	49.2 [241]	121 [241]
Polyester of hydroxybenzoic and Hydroxynapthoic acids	$C_{39}H_{22}O_{10}$	11.1 [236]	40.6 [241]	164 [241]

(continued)

(continued)

Name	Elemental composition in repeat unit composition	Total heat release	Char yield	heat release capacity
LaRC-TOR	$C_{44}H_{29}N_4O_3P$	11.7 [236]	63.0 [241]	135 [241]
LaRC-CP2	$C_{37}H_{18}N_2O_6F_6$	3.4 [236]	57.0 [241]	14 [241]
LaRC-CP1	$C_{46}H_{22}N_2O_6F_{12}$	2.9 [236]	52.0 [241]	13 [241]
PHA-1	$C_{20}O_4N_2H_{14}$	10.0 [237]	56.0 [237]	42 [237]
PHA-5	$C_{23}N_2O_3H_{12}$	3.0 [237]	36.0 [237]	8 [237]
PHA-7	$C_{22}N_2O_4H_{18}$	17.0 [237]	43.0 [237]	130 [237]
PHA-11	$C_{44}P_2O_6N_2H_{32}$	21.0 [237]	32.0 [237]	210 [237]
PHA-12	$C_{28}N_2P_2O_{10}H_{32}$	9.0 [237]	41.0 [237]	73 [237]
BEDB/4,4'-DDS	$C_{34}H_{32}N_2O_7S$	17.2 [238]	30.0 [238]	420 [238]
BEDB/3,3'-DDS	$C_{34}H_{32}N_2O_7S$	17.2 [238]	33.0 [238]	429 [238]
BEDB/m-PDA	$C_{26}H_{28}N_2O_5$	19.7 [238]	29.0 [238]	391 [238]
BEDB/2,4-DT	$C_{27}H_{30}N_2O_5$	17.0 [238]	33.0 [238]	372 [238]
BEDB/2,6-DT	$C_{27}H_{30}N_2O_5$	17.9 [238]	32.0 [238]	530 [238]
BEDB/2,3-DT	$C_{27}H_{30}N_2O_5$	20.0 [238]	22.0 [238]	326 [238]
BEDB/3,4-DT	$C_{27}H_{30}N_2O_5$	20.3 [238]	20.0 [238]	415 [238]
BEDB/4-CmP	$C_{26}H_{27}N_2O_5Cl$	15.3 [238]	36.0 [238]	169 [238]
BEDB/5-CmP	$C_{26}H_{27}N_2O_5Cl$	15.3 [238]	34.0 [238]	292 [238]
BEDB/4-CoP	$C_{26}H_{27}N_2O_5Cl$	17.6 [238]	21.0 [238]	389 [238]
BEDB/3-3'-DMB	$C_{30}H_{36}N_2O_5$	17.2 [238]	41.0 [238]	400 [238]
BEDB/3-3'-DMoB	$C_{30}H_{36}N_2O_7$	14.8 [238]	42.0 [238]	288 [238]
EBPA/DDM	$C_{34}H_{38}N_2O_4$	25.8 [238]	16.0 [238]	737 [238]
EBPA/DDS	$C_{33}H_{36}N_2O_6S$	25.3 [238]	12.0 [238]	513 [238]
EBPA/m-PDA	$C_{27}H_{35}N_2O_4$	25.8 [238]	14.0 [238]	761 [238]
ETBBA/DDM	$C_{34}H_{34}N_2O_4Br_4$	4.5 [238]	24.0 [238]	308 [238]
ETBBA/DDS	$C_{33}H_{32}N_2O_6Br_4S$	4.8 [238]	23.0 [238]	443 [238]
ETBBA/m-PMA	$C_{27}H_{24}Br_4N_4$	5.0 [238]	23.0 [238]	238 [238]
BPA Polyarylate	$C_{22}O_5Cl_2H_{12}$	18.0 [239]	27.0 [239]	360 [239]
Chalcon II	$C_{25}O_6H_{18}$	10.0 [239]	41.0 [239]	110 [239]
BHDB-sulfone (1)	$C_{26}H_{18}O_5S$	10.0 [240]	-	-
BHDB-sulfoxide (2)	$C_{26}H_{18}O_4S$	6.5 [240]	-	-
BHDB/biphenyl (9)	$C_{25}H_{17}O_{4.5}S$	8.0 [240]	-	-
BHDB/sufide (10)	$C_{25}H_{17}O_{4.5}S_{1.5}$	10.7 [240]	-	-
PPSU	$C_{24}H_{16}O_4S$	13.5 [267]	-	-
BDHB Acrylate	$C_{22}H_{14}O_5$	6.0 [267]	-	-
BDHB Phosphinate	$C_{20}H_{14}O_4P$	9.5 [267]	-	-
BHDB poly(arylate-co-phos-phonate) (1:1)	$C_{42}H_{28}O_9P$	5.1 [267]	-	-
Polyester carbonate (white)	$C_{37}H_{26}O_{10}$	10.9 [267]	-	-
Polyester carbonate (gray)	$C_{37}H_{26}O_{10}$	10.4 [267]	-	-
Polyester carbonate (clear)	$C_{37}H_{26}O_{10}$	10.3 [267]	-	-

References

[1] ASTM, Annual book of ASTM standards, American Society for Testing and Materials, Philadelphia; 2002.

[2] S. Mannan, Lees' Loss prevention in the process industries: Hazard identification, assessment and control, Butterworth-Heinemann, Oxford, 2012.

[3] R. King, Safety in the process industries, Butterworth-Heinemann, London, 1990.

[4] Z. Liu, A. Kim, D. Carpenter, Extinguishment of large cooking oil pool fires by the use of water mist system, in: Combustion Institute/Canada Section, Spring Technical Meeting, 2004, pp. 1–6.

[5] D.S.J. Jones, P.R. Pujadó, Handbook of petroleum processing, Springer, Dordrecht, 2006.

[6] D. Kong, D.J. am Ende, S.J. Brenek, N.P. Weston, Determination of flash point in air and pure oxygen using an equilibrium closed bomb apparatus, Journal of Hazardous Materials, 102 (2003) 155–165.

[7] R.C. Lance, A.J. Barnard, J.E. Hooyman, Measurement of flash points: apparatus, methodology, applications, Journal of Hazardous Materials, 3 (1979) 107–119.

[8] A. D240-09, Standard test method for heat of combustion of liquid hydrocarbon fuels by bomb calorimeter, in: ASTM International, West Conshohocken, PA, 2009.

[9] A. D93, Standard Test Methods for flash point by Pensky-Martens closed cup tester, in: ASTM International, West Conshohocken, PA, 2002.

[10] E.W. White, R.G. Montemayor, The practice of flash point determination: A laboratory resource (MNL72), ASTM, West Conshohocken, PA., USA, 2013.

[11] A. D92, Standard test method for flash and fire points by cleveland open cup tester, in: ASTM International, West Conshohocken, PA, 2005.

[12] J. Rowley, D. Freeman, R. Rowley, J. Oscarson, N. Giles, W. Wilding, Flash point: Evaluation, experimentation and estimation, International Journal of Thermophysics, 31 (2010) 875–887.

[13] H. Ishida, A. Iwama, Some critical discussions on flash and fire points of liquid fuels, in: Fire Safety Science–Proceedings of the first international symposium, 1986, pp. 217–226.

[14] C. Ding, W. Yao, Y. Tang, J. Rong, D. Zhou, J. Wang, Experimental study of the flash point of flammable liquids under different altitudes in Tibet plateau, Fire and Materials, 38 (2014) 241–246.

[15] C. Ding, W. Yao, D. Zhou, J. Rong, Y. Zhang, W. Zhang, J. Wang, Experimental study and hazard analysis on the flash point of flammable liquids at high altitudes, Journal of Fire Sciences, 31 (2013) 469–477.

[16] L.G. Britton, K.L. Cashdollar, W. Fenlon, D. Frurip, J. Going, B.K. Harrison, J. Niemeier, E.A. Ural, The role of ASTM E27 methods in hazard assessment part II: Flammability and ignitability, Process Safety Progress, 24 (2005) 12–28.

[17] A.R. Katritzky, R. Petrukhin, R. Jain, M. Karelson, QSPR analysis of flash points, Journal of Chemical Information and Computer Sciences, 41 (2001) 1521–1530.

[18] T. Suzuki, K. Ohtaguchi, K. Koide, A method for estimating flash points of organic compounds from molecular structures, Journal of Chemical Engineering of Japan, 24 (1991) 258–261.

[19] L. Catoire, V. Naudet, A unique equation to estimate flash points of selected pure liquids application to the correction of probably erroneous flash point values, Journal of Physical and Chemical Reference Data, 33 (2004) 1083–1111.

[20] M. Vidal, W. Rogers, J. Holste, M. Mannan, A review of estimation methods for flash points and flammability limits, Process Safety Progress, 23 (2004) 47–55.

[21] X. Liu, Z. Liu, Research progress on flash point prediction, Journal of Chemical & Engineering Data, 55 (2010) 2943–2950.

[22] R.W. Prugh, Estimation of flash point temperature, Journal of Chemical Education, 50 (1973) A85.

https://doi.org/10.1515/9783110572223-010

[23] A. Fujii, E.R. Hermann, Correlation between flash points and vapor pressures of organic compounds, Journal of Safety Research, 13 (1982) 163–175.

[24] G. Patil, Estimation of flash point, Fire and Materials, 12 (1988) 127–131.

[25] K. Satyanarayana, M. Kakati, Note: Correlation of flash points, Fire and Materials, 15 (1991) 97–100.

[26] K. Satyanarayana, P. Rao, Improved equation to estimate flash points of organic compounds, Journal of Hazardous Materials, 32 (1992) 81–85.

[27] E. Metcalfe, A. Metcalfe, Communication: On the correlation of flash points, Fire and Materials, 16 (1992) 153–154.

[28] F.-Y. Hshieh, Note: correlation of closed-cup flash points with normal boiling points for silicone and general organic compounds, Fire and Materials, 21 (1997) 277–282.

[29] L. Catoire, S. Paulmier, V. Naudet, Experimental determination and estimation of closed cup flash points of mixtures of flammable solvents, Process Safety Progress, 25 (2006) 33–39.

[30] E.M. Valenzuela, R. Vázquez-Román, S. Patel, M.S. Mannan, Prediction models for the flash point of pure components, Journal of Loss Prevention in the Process Industries, 24 (2011) 753–757.

[31] D.A. Saldana, L. Starck, P. Mougin, B. Rousseau, B. Creton, Prediction of flash points for fuel mixtures using machine learning and a novel equation, Energy & Fuels, 27 (2013) 3811–3820.

[32] H.-J. Liaw, T.-P. Tsai, Flash-point estimation for binary partially miscible mixtures of flammable solvents by UNIFAC group contribution methods, Fluid Phase Equilibria, 375 (2014) 275–285.

[33] L.Y. Phoon, A.A. Mustaffa, H. Hashim, R. Mat, A Review of Flash Point Prediction Models for Flammable Liquid Mixtures, Industrial & Engineering Chemistry Research, 53 (2014) 12553–12565.

[34] F. Gharagheizi, M.H. Keshavarz, M. Sattari, A simple accurate model for prediction of flash point temperature of pure compounds, Journal of Thermal Analysis and Calorimetry, 110 (2012) 1005–1012.

[35] D. Mathieu, Flash points of organosilicon compounds: How data for alkanes combined with custom additive fragments can expedite the development of predictive models, Industrial & Engineering Chemistry Research, 51 (2012) 14309–14315.

[36] F.A. Carroll, C.-Y. Lin, F.H. Quina, Simple method to evaluate and to predict flash points of organic compounds, Industrial & Engineering Chemistry Research, 50 (2011) 4796–4800.

[37] M. Riazi, T. Daubert, Predicting flash and pour points, Hydrocarbon Processing, 66 (1987) 81–83.

[38] F. Gharagheizi, A. Eslamimanesh, A.H. Mohammadi, D. Richon, Empirical method for representing the flash-point temperature of pure compounds, Industrial & Engineering Chemistry Research, 50 (2011) 5877–5880.

[39] J.R. Rowley, R.L. Rowley, W.V. Wilding, Prediction of pure-component flash points for organic compounds, Fire and Materials, 35 (2011) 343–351.

[40] J. Rowley, R. Rowley, W. Wilding, Estimation of the flash point of pure organic chemicals from structural contributions, Process Safety Progress, 29 (2010) 353–358.

[41] F. Gharagheizi, P. Ilani-Kashkouli, N. Farahani, A.H. Mohammadi, Gene expression programming strategy for estimation of flash point temperature of non-electrolyte organic compounds, Fluid Phase Equilibria, 329 (2012) 71–77.

[42] B.E. Poling, J.M. Prausnitz, O.C. John Paul, R.C. Reid, The properties of gases and liquids, 5th ed., McGraw-Hill, New York, 2001.

[43] Project 801, Evaluated process design data of the Design Institute for Physical Properties (DIPPR), AIChE, New York, 2014.

[44] A. Alibakhshi, H. Mirshahvalad, S. Alibakhshi, A modified group contribution method for accurate prediction of flash points of pure organic compounds, Industrial & Engineering Chemistry Research, 54 (2015) 11230–11235.

[45] T.A. Albahri, MNLR and ANN structural group contribution methods for predicting the flash point temperature of pure compounds in the transportation fuels range, Process Safety and Environmental Protection, 93 (2015) 182–191.

[46] J.A. Lazzús, Prediction of flash point temperature of organic compounds using a hybrid method of group contribution+ neural network+ particle swarm optimization, Chinese Journal of Chemical Engineering, 18 (2010) 817–823.

[47] A. Alibakhshi, H. Mirshahvalad, S. Alibakhshi, Prediction of flash points of pure organic compounds: Evaluation of the DIPPR database, Process Safety and Environmental Protection, 105 (2017) 127–133.

[48] F.Z. Serat, A.M. Benkouider, A. Yahiaoui, F. Bagui, Nonlinear group contribution model for the prediction of flash points using normal boiling points, Fluid Phase Equilibria, (2017).

[49] S.W. Benson, F. Cruickshank, D. Golden, G.R. Haugen, H. O'Neal, A. Rodgers, R. Shaw, R. Walsh, Additivity rules for the estimation of thermochemical properties, Chemical Reviews, 69 (1969) 279–324.

[50] K. Argoub, A.M. Benkouider, A. Yahiaoui, R. Kessas, S. Guella, F. Bagui, Prediction of standard enthalpy of formation in the solid state by a third-order group contribution method, Fluid Phase Equilibria, 380 (2014) 121–127.

[51] M. Kamalvand, M.H. Keshavarz, M. Jafari, Prediction of the strength of energetic materials using the condensed and gas phase heats of formation, Propellants, Explosives, Pyrotechnics, 40 (2015) 551–557.

[52] S.H. Yalkowsky, Carnelley's rule and the prediction of melting point, Journal of Pharmaceutical Sciences, 103 (2014) 2629–2634.

[53] A. Jain, G. Yang, S.H. Yalkowsky, Estimation of melting points of organic compounds, Industrial & Engineering Chemistry Research, 43 (2004) 7618–7621.

[54] F. Gharagheizi, P. Ilani-Kashkouli, W.E. Acree Jr, A.H. Mohammadi, D. Ramjugernath, A group contribution model for determining the sublimation enthalpy of organic compounds at the standard reference temperature of 298 K, Fluid Phase Equilibria, 354 (2013) 265–285.

[55] T.A. Albahri, Flammability characteristics of pure hydrocarbons, Chemical Engineering Science, 58 (2003) 3629–3641.

[56] Y. Pan, J. Jiang, Z. Wang, Quantitative structure–property relationship studies for predicting flash points of alkanes using group bond contribution method with back-propagation neural network, Journal of Hazardous Materials, 147 (2007) 424–430.

[57] F. Gharagheizi, R.F. Alamdari, M.T. Angaji, A new neural network– group contribution method for estimation of flash point temperature of pure components, Energy & Fuels, 22 (2008) 1628–1635.

[58] G.-B. Wang, C.-C. Chen, H.-J. Liaw, Y.-J. Tsai, Prediction of flash points of organosilicon compounds by structure group contribution approach, Industrial & Engineering Chemistry Research, 50 (2011) 12790–12796.

[59] API, Technical data book-petroleum refining, 6th ed. ed., The American Petroleum Institute, Washington, DC, 1987.

[60] J. Tetteh, T. Suzuki, E. Metcalfe, S. Howells, Quantitative structure– property relationships for the estimation of boiling point and flash point using a radial basis function neural network, Journal of Chemical Information and Computer Sciences, 39 (1999) 491–507.

[61] A.R. Katritzky, I.B. Stoyanova-Slavova, D.A. Dobchev, M. Karelson, QSPR modeling of flash points: An update, Journal of Molecular Graphics and Modelling, 26 (2007) 529–536.

[62] F. Gharagheizi, R.F. Alamdari, Prediction of flash point temperature of pure components using a quantitative structure–property relationship model, Molecular Informatics, 27 (2008) 679–683.

[63] A. Mauri, V. Consonni, M. Pavan, R. Todeschini, Dragon software: An easy approach to molecular descriptor calculations, MATCH-Communications in Mathematical and in Computer Chemistry, 56 (2006) 237–248.

[64] M.H. Keshavarz, M. Ghanbarzadeh, Simple method for reliable predicting flash points of unsaturated hydrocarbons, Journal of Hazardous Materials, 193 (2011) 335–341.

[65] M.H. Keshavarz, Estimation of the flash points of saturated and unsaturated hydrocarbons, Indian Journal of Engineering & Materials Sciences, 19 (2012) 269–278.

[66] M.H. Keshavarz, S. Moradi, A.R. Madram, H.R. Pouretedal, K. Esmailpour, A. Shokrolahi, Reliable method for prediction of the flash point of various classes of amines on the basis of some molecular moieties for safety measures in industrial processes, Journal of Loss Prevention in the Process Industries, 26 (2013) 650–659.

[67] M.H. Keshavarz, M. Jafari, M. Kamalvand, A. Karami, Z. Keshavarz, A. Zamani, S. Rajaee, A simple and reliable method for prediction of flash point of alcohols based on their elemental composition and structural parameters, Process Safety and Environmental Protection, 102 (2016) 1–8.

[68] N. Zohari, M.M.E. Qhomi, Two Reliable Simple Relationships between flash points of hydrocarbon kerosene fuels and their molecular structures, Zeitschrift für anorganische und allgemeine Chemie, 643 (2017) 985–992.

[69] M.H. Keshavarz, H. Motamedoshariati, M. Ghanbarzadeh, A reliable simple method for prediction of the flash points of saturated hydrocarbons in order to improve their safety, Chemistry, 20 (2011) 58–75.

[70] *Hazardous Chemical Database (http://ull.chemistry.uakron.edu/erd/)* The University of Akron maintains a database that gives the set of physical properties commonly found on MSDS for about 4,000 compounds.

[71] R.M. Stephenson, Flash points of organic and organometallic compounds, Elsevier, New York, 1987.

[72] W.A. Affens, G.W. McLaren, Flammability properties of hydrocarbon solutions in air, Journal of Chemical and Engineering Data, 17 (1972) 482–488.

[73] L. Catoire, S. Paulmier, V. Naudet, Estimation of closed cup flash points of combustible solvent blends, Journal of Physical and Chemical Reference Data, 35 (2006) 9–14.

[74] R.W. Garland, M.O. Malcolm, Evaluating vent manifold inerting requirements: Flash point modeling for organic acid-water mixtures, Process Safety Progress, 21 (2002) 254–260.

[75] J. Gmehling, P. Rasmussen, Flash points of flammable liquid mixtures using UNIFAC, Industrial & Engineering Chemistry Fundamentals, 21 (1982) 186–188.

[76] H.-J. Liaw, Y.-Y. Chiu, A general model for predicting the flash point of miscible mixtures, Journal of Hazardous Materials, 137 (2006) 38–46.

[77] H.-J. Liaw, Y.-Y. Chiu, The prediction of the flash point for binary aqueous-organic solutions, Journal of Hazardous Materials, 101 (2003) 83–106.

[78] H.-J. Liaw, T.-A. Wang, A non-ideal model for predicting the effect of dissolved salt on the flash point of solvent mixtures, Journal of Hazardous Materials, 141 (2007) 193–201.

[79] H.-J. Liaw, Y.-H. Lee, C.-L. Tang, H.-H. Hsu, J.-H. Liu, A mathematical model for predicting the flash point of binary solutions, Journal of Loss Prevention in the Process Industries, 15 (2002) 429–438.

[80] H.-J. Liaw, C.-L. Tang, J.-S. Lai, A model for predicting the flash point of ternary flammable solutions of liquid, Combustion and Flame, 138 (2004) 308–319.

[81] S. Lee, D.-M. Ha, The lower flash points of binary systems containing non-flammable component, Korean Journal of Chemical Engineering, 20 (2003) 799–802.

[82] D. White, C.L. Beyler, C. Fulper, J. Leonard, Flame spread on aviation fuels, Fire Safety Journal, 28 (1997) 1–31.

[83] H. Le Chatelier, Estimation of firedamp by flammability limits, Annals of Mines, 19 (1891) 388–395.

[84] C.V. Mashuga, D.A. Crowl, Derivation of Le Chatelier's mixing rule for flammable limits, Process Safety Progress, 19 (2000) 112–117.

[85] H.-J. Liaw, S.-C. Lin, Binary mixtures exhibiting maximum flash-point behavior, Journal of Hazardous Materials, 140 (2007) 155–164.

[86] H.-J. Liaw, T.-P. Lee, J.-S. Tsai, W.-H. Hsiao, M.-H. Chen, T.-T. Hsu, Binary liquid solutions exhibiting minimum flash-point behavior, Journal of loss Prevention in the Process Industries, 16 (2003) 173–186.

[87] H.-J. Liaw, C.-T. Chen, V. Gerbaud, Flash-point prediction for binary partially miscible aqueous–organic mixtures, Chemical Engineering Science, 63 (2008) 4543–4554.

[88] H.-J. Liaw, V. Gerbaud, Y.-H. Li, Prediction of miscible mixtures flash-point from UNIFAC group contribution methods, Fluid Phase Equilibria, 300 (2011) 70–82.

[89] A. Fredenslund, R.L. Jones, J.M. Prausnitz, Group-contribution estimation of activity coefficients in nonideal liquid mixtures, AIChE Journal, 21 (1975) 1086–1099.

[90] A. Fredenslund, J. Gmehling, M.L. Michelsen, P. Rasmussen, J.M. Prausnitz, Computerized design of multicomponent distillation columns using the UNIFAC group contribution method for calculation of activity coefficients, Industrial & Engineering Chemistry Process Design and Development, 16 (1977) 450–462.

[91] B.L. Larsen, P. Rasmussen, A. Fredenslund, A modified UNIFAC group-contribution model for prediction of phase equilibria and heats of mixing, Industrial & Engineering Chemistry Research, 26 (1987) 2274–2286.

[92] J. Gmehling, J. Li, M. Schiller, A modified UNIFAC model. 2. Present parameter matrix and results for different thermodynamic properties, Industrial & Engineering Chemistry Research, 32 (1993) 178–193.

[93] H.H. Hooper, S. Michel, J.M. Prausnitz, Correlation of liquid-liquid equilibria for some water-organic liquid systems in the region 20–250°C, Industrial & Engineering Chemistry Research, 27 (1988) 2182–2187.

[94] M. Vidal, W. Rogers, M. Mannan, Prediction of minimum flash point behaviour for binary mixtures, Process Safety and Environmental Protection, 84 (2006) 1–9.

[95] U. Weidlich, J. Gmehling, A modified UNIFAC model. I: Prediction of VLE, hE, and γ∞, Industrial & Engineering Chemistry Research, 26 (1987) 1372–1381.

[96] H.-J. Liaw, T.-P. Tsai, Flash points of partially miscible aqueous–organic mixtures predicted by UNIFAC group contribution methods, Fluid Phase Equilibria, 345 (2013) 45–59.

[97] H.-J. Liaw, H.-Y. Chen, Study of two different types of minimum flash-point behavior for ternary mixtures, Industrial & Engineering Chemistry Research, 52 (2013) 7579–7585.

[98] Y. Pan, J. Cheng, X. Song, G. Li, L. Ding, J. Jiang, Flash points measurements and prediction for binary miscible mixtures, Journal of Loss Prevention in the Process Industries, 34 (2015) 56–64.

[99] B.E. Poling, J.M. Prausnitz, J.P. O'Connell, The properties of gases and liquids, Mcgraw-Hill, New York, 2001.

[100] T. Boublík, V. Fried, E. Hála: The vapour pressures of pure substances-Selected values of the temperature dependence of the vapour pressures of some pure substances in the normal and low pressure region (second revised edition), Vol. 17 aus: Physical Sciences Data, Elsevier Science Publishers, Amsterdam, Oxford, New York, Tokyo 1984.

[101] W.M. Haynes, CRC handbook of chemistry and physics, CRC Press, Boca Raton, 2014.

[102] *NIST Chemistry WebBook (http://webbook.nist.gov/chemistry/)*

[103] H. Pouretedal, M. H. Keshavarz, Prediction of toxicity of nitroaromatic compounds through their molecular structures, Journal of the Iranian Chemical Society, 8 (2011) 78–89.

[104] M.H. Keshavarz, A. Ramadan, A. Mousaviazar, A. Zali, K. Esmaeilpour, F. Atabaki, A. Shokrolahi, Reducing dangerous effects of unsymmetrical dimethylhydrazine as a liquid

propellant by addition of hydroxyethylhydrazine – Part I: Physical properties, Journal of Energetic Materials, 29 (2011) 46–60.

[105] M.H. Keshavarz, A. Ramadan, A. Mousaviazar, A. Zali, A. Shokrolahi, Reducing dangerous effects of unsymmetrical dimethylhydrazine as a liquid propellant by addition of hydroxyethylhydrazine – Part II: Performance with several oxidizers, Journal of Energetic Materials, 29 (2011) 228–240.

[106] S.G. Pakdehi, S. Ajdari, A. Hashemi, M.H. Keshavarz, Performance evaluation of liquid fuel 2-dimethyl amino ethyl azide (DMAZ) with liquid oxidizers, Journal of Energetic Materials, 33 (2015) 17–23.

[107] S.G. Pakdehi, S. Rezaei, H. Motamedoshariati, M.H. Keshavarz, Sensitivity of dimethyl amino ethyl azide (DMAZ) as a non-carcinogenic and high performance fuel to some external stimuli, Journal of Loss Prevention in the Process Industries, 29 (2014) 277–282.

[108] S.G. Pakdehi, M.H. Keshavarz, M. Akbari, M. Ghorbani, Assessment of physico-thermal properties, combustion performance, and ignition delay time of dimethyl amino ethanol as a novel liquid fuel, Propellants, Explosives, Pyrotechnics, 42 (2017) 423–429.

[109] T.A. Albahri, R.S. George, Artificial neural network investigation of the structural group contribution method for predicting pure components auto ignition temperature, Industrial & Engineering Chemistry Research, 42 (2003) 5708–5714.

[110] C.-C. Chen, H.-J. Liaw, Y.-Y. Kuo, Prediction of autoignition temperatures of organic compounds by the structural group contribution approach, Journal of Hazardous Materials, 162 (2009) 746–762.

[111] D.E. Swarts, M. Orchin, Spontaneous ignition temperature of hydrocarbons, Industrial & Engineering Chemistry, 49 (1957) 432–436.

[112] M.H. Keshavarz, F. Gharagheizi, M. Ghanbarzadeh, A simple correlation for prediction of autoignition temperature of various classes of hydrocarbons, Journal of the Iranian Chemical Society, 10 (2013) 545–557.

[113] A. E659-78, Standard Test Method for Autoignition Temperature of Liquid Chemicals, in: ASTM International, West Conshohocken, PA, 2005.

[114] T. Suzuki, K. Ohtaguchi, K. Koide, Correlation and prediction of autoignition temperatures of hydrocarbons using molecular properties, Journal of Chemical Engineering of Japan, 25 (1992) 606–608.

[115] C.-C. Huang, T.-S. Wu, A simple method for estimating the autoignition temperature of solid energetic materials with a single non-isothermal DSC or DTA curve, Thermochimica Acta, 239 (1994) 105–114.

[116] T. Suzuki, Quantitative structure – property relationships for auto-ignition temperatures of organic compounds, Fire and Materials, 18 (1994) 81–88.

[117] J. Tetteh, E. Metcalfe, S.L. Howells, Optimisation of radial basis and backpropagation neural networks for modelling auto-ignition temperature by quantitative-structure property relationships, Chemometrics and Intelligent Laboratory Systems, 32 (1996) 177–191.

[118] B.E. Mitchell, P.C. Jurs, Prediction of autoignition temperatures of organic compounds from molecular structure, Journal of Chemical Information and Computer Sciences, 37 (1997) 538–547.

[119] Y.S. Kim, S.K. Lee, J.H. Kim, J.S. Kim, K.T. No, Prediction of autoignition temperatures (AITs) for hydrocarbons and compounds containing heteroatoms by the quantitative structure–property relationship, Journal of the Chemical Society, Perkin Transactions, 2 (2002) 2087–2092.

[120] Y. Pan, J. Jiang, R. Wang, H. Cao, Advantages of support vector machine in QSPR studies for predicting auto-ignition temperatures of organic compounds, Chemometrics and Intelligent Laboratory Systems, 92 (2008) 169–178.

[121] Y. Pan, J. Jiang, R. Wang, H. Cao, J. Zhao, Prediction of auto-ignition temperatures of hydrocarbons by neural network based on atom-type electrotopological-state indices, Journal of Hazardous Materials, 157 (2008) 510–517.

[122] Y. Pan, J. Jiang, R. Wang, H. Cao, Y. Cui, Predicting the auto-ignition temperatures of organic compounds from molecular structure using support vector machine, Journal of Hazardous Materials, 164 (2009) 1242–1249.

[123] Y. Pan, J. Jiang, X. Ding, R. Wang, J. Jiang, Prediction of flammability characteristics of pure hydrocarbons from molecular structures, AIChE Journal, 56 (2010) 690–701.

[124] F. Gharagheizi, An accurate model for prediction of autoignition temperature of pure compounds, Journal of Hazardous Materials, 189 (2011) 211–221.

[125] J.A. Lazzús, Autoignition temperature prediction using an artificial neural network with particle swarm optimization, International Journal of Thermophysics, 32 (2011) 957–973.

[126] F.-Y. Tsai, C.-C. Chen, H.-J. Liaw, A model for predicting the auto-ignition temperature using quantitative structure property relationship approach, Procedia Engineering, 45 (2012) 512–517.

[127] M. Bagheri, T.N.G. Borhani, G. Zahedi, Estimation of flash point and autoignition temperature of organic sulfur chemicals, Energy Conversion and Management, 58 (2012) 185–196.

[128] J.A. Lazzús, Neural network-particle swarm modeling to predict thermal properties, Mathematical and Computer Modelling, 57 (2013) 2408–2418.

[129] T.N.G. Borhani, A. Afzali, M. Bagheri, QSPR estimation of the auto-ignition temperature for pure hydrocarbons, Process Safety and Environmental Protection, 103 (2016) 115–125.

[130] L.M. Egolf, P.C. Jurs, Estimation of autoignition temperatures of hydrocarbons, alcohols, and esters from molecular structure, Industrial & Engineering Chemistry Research, 31 (1992) 1798–1807.

[131] M.H. Keshavarz, M. Jafari, K. Esmaeilpour, M. Samiee, New and reliable model for prediction of autoignition temperature of organic compounds containing energetic groups, Process Safety and Environmental Protection, 113 (2018) 491–497.

[132] M.H. Keshavarz, A new general correlation for predicting impact sensitivity of energetic compounds, Propellants, Explosives, Pyrotechnics, 38 (2013) 754–760.

[133] M.H. Keshavarz, H. Motamedoshariati, H.R. Pouretedal, M.K. Tehrani, A. Semnani, Prediction of shock sensitivity of explosives based on small-scale gap test, Journal of Hazardous Materials, 145 (2007) 109–112.

[134] M.H. Keshavarz, H.R. Pouretedal, A. Semnani, Reliable prediction of electric spark sensitivity of nitramines: A general correlation with detonation pressure, Journal of Hazardous Materials, 167 (2009) 461–466.

[135] R.P. Pohanish, S.A. Greene, Wiley guide to chemical incompatibilities, John Wiley & Sons, Hoboken, 2009.

[136] R.P. Pohanish, Sittig's handbook of toxic and hazardous chemicals and carcinogens, William Andrew, Amsterdam, 2012.

[137] N.P. Cheremisinoff, Handbook of industrial toxicology and hazardous materials, CRC Press, 1999.

[138] M. Sheldon, A study of the flammability limits of gases and vapors, Fire Prevention, 174 (1984) 23–31.

[139] A.E. 681, Standard test method for concentration limits of flammability of chemicals (Vapors and gases), in: ASTM International, Philadelphia, Pennsylvania, 2004.

[140] W.A. Affens, Flammability properties of hydrocarbon fuels. Interrelations of flammability properties of n-alkanes in air, Journal of Chemical and Engineering Data, 11 (1966) 197–202.

[141] V. Babrauskas, Ignition handbook, Fire Science Publishers, Issaquah, WA, 2003.

[142] L.G. Britton, Using heats of oxidation to evaluate flammability hazards, Process Safety Progress, 21 (2002) 31–54.

[143] L.G. Britton, D.J. Frurip, Further uses of the heat of oxidation in chemical hazard assessment, Process Safety Progress, 22 (2003) 1–19.

[144] D. Dalmazzone, J.-C. Laforest, J. Petit, Application of thermochemical energy hazard criteria to the prediction of lower flammability limits of hydrocarbons in air, Oil & Gas Science and Technology, 56 (2001) 365–372.

[145] R. Ervin, M. Palucis, T. Glowienka, V. Van Brunt, W. Chastain, R. Kline, P. Lodal, Using the adiabatic flame temperature to predict the flammability of lower alkanes, carboxylic acids and acetates, in: Conference Proceedings/National Meeting AIChe, 2005.

[146] F. Funk, Calculation of the lower explosivity limit of combustible gases and vapors, Chemical Technology, 26 (1974) 779–780.

[147] C. Hilado, A method for estimating limits of flammability(organic compound hazards), Journal of Fire and Flammability, 6 (1975) 130–139.

[148] F.-Y. Hshieh, Predicting heats of combustion and lower flammability limits of organosilicon compounds, Fire and Materials, 23 (1999) 79–89.

[149] S. Kondo, A. Takahashi, K. Tokuhashi, A. Sekiya, RF number as a new index for assessing combustion hazard of flammable gases, Journal of Hazardous Materials, 93 (2002) 259–267.

[150] S. Kondo, Y. Urano, K. Tokuhashi, A. Takahashi, K. Tanaka, Prediction of flammability of gases by using F-number analysis, Journal of Hazardous Materials, 82 (2001) 113–128.

[151] C.V. Mashuga, D.A. Crowl, Flammability zone prediction using calculated adiabatic flame temperatures, Process Safety Progress, 18 (1999) 127–134.

[152] G. Melhem, A detailed method for estimating mixture flammability limits using chemical equilibrium, Process Safety Progress, 16 (1997) 203–218.

[153] M. Miloshev, D. Vulchev, Z. Zdravchev, Relation between the concentration limits of flammability and physicochemical indexes of hydrocarbons, Godishnik na Visshiya Khimiko-Tekhnologicheski Institut, Sofiya, 27 (1982) 92–107.

[154] W. Möller, P. Schulz, T. Redeker, Procedure for estimating flash points and lower explosions limits, Physikalisch-Technische Bundestanstalt Thermodynamik, PTB Report W-55, (1993).

[155] W.H. Seaton, Group contribution method for predicting the lower and the upper flammable limits of vapors in air, Journal of Hazardous Materials, 27 (1991) 169–185.

[156] Y.N. Shebeko, W. Fan, I. Bolodian, V.Y. Navzenya, An analytical evaluation of flammability limits of gaseous mixtures of combustible–oxidizer–diluent, Fire Safety Journal, 37 (2002) 549–568.

[157] Y.N. Shebeko, A. Ivanov, T. Dmitrieva, Methods of calculation of lower concentration limits of combustion of gases and vapors in air, The Soviet Chemical Industry, 15 (1983) 311–314.

[158] A. Shimy, Calculating flammability characteristics of hydrocarbons and alcohols, Fire Technology, 6 (1970) 135–139.

[159] N. Solov'ev, A. Baratov, Lower limit of flammability of hydrocarbon–air mixtures as a function of the molecular structure of the combustible component, Russian Journal of Physical Chemistry, 34 (1960) 1661–1670.

[160] T. Suzuki, Empirical relationship between lower flammability limits and standard enthalpies of combustion of organic compounds, Fire and Materials, 18 (1994) 333–336.

[161] T. Suzuki, M. Ishida, Neural network techniques applied to predict flammability limits of organic compounds, Fire and Materials, 19 (1995) 179–189.

[162] M. Vidal, W. Wong, W. Rogers, M.S. Mannan, Evaluation of lower flammability limits of fuel–air–diluent mixtures using calculated adiabatic flame temperatures, Journal of Hazardous Materials, 130 (2006) 21–27.

[163] J.R. Rowley, R.L. Rowley, W.V. Wilding, Experimental determination and re-examination of the effect of initial temperature on the lower flammability limit of pure liquids, Journal of Chemical & Engineering Data, 55 (2010) 3063–3067.

[164] L. Catoire, V. Naudet, Estimation of temperature-dependent lower flammability limit of pure organic compounds in air at atmospheric pressure, Process Safety Progress, 24 (2005) 130–137.

[165] R. Rowley, W. Wilding, J. Oscarson, N. Giles, DIPPR® Data compilation of pure chemical properties, design institute for physical properties, AIChE, New York, 2010.

[166] F. Gharagheizi, Quantitative structure– property relationship for prediction of the lower flammability limit of pure compounds, Energy & Fuels, 22 (2008) 3037–3039.

[167] F. Gharagheizi, A new group contribution-based model for estimation of lower flammability limit of pure compounds, Journal of Hazardous Materials, 170 (2009) 595–604.

[168] J. Rowley, R. Rowley, W. Wilding, Estimation of the lower flammability limit of organic compounds as a function of temperature, Journal of Hazardous Materials, 186 (2011) 551–557.

[169] F. Gharagheizi, Prediction of upper flammability limit percent of pure compounds from their molecular structures, Journal of Hazardous Materials, 167 (2009) 507–510.

[170] F. Gharagheizi, Chemical structure-based model for estimation of the upper flammability limit of pure compounds, Energy & Fuels, 24 (2010) 3867–3871.

[171] F. Gharagheizi, P. Ilani-Kashkouli, A.H. Mohammadi, Corresponding states method for estimation of upper flammability limit temperature of chemical compounds, Industrial & Engineering Chemistry Research, 51 (2012) 6265–6269.

[172] J.A. Lazzús, Neural network/particle swarm method to predict flammability limits in air of organic compounds, Thermochimica Acta, 512 (2011) 150–156.

[173] J.A. Lazzús, Prediction of flammability limit temperatures from molecular structures using a neural network–particle swarm algorithm, Journal of the Taiwan Institute of Chemical Engineers, 42 (2011) 447–453.

[174] F. Gharagheizi, New neural network group contribution model for estimation of lower flammability limit temperature of pure compounds, Industrial & Engineering Chemistry Research, 48 (2009) 7406–7416.

[175] M. Bagheri, M. Rajabi, M. Mirbagheri, M. Amin, BPSO-MLR and ANFIS based modeling of lower flammability limit, Journal of Loss Prevention in the Process Industries, 25 (2012) 373–382.

[176] T.A. Albahri, Prediction of the lower flammability limit percent in air of pure compounds from their molecular structures, Fire Safety Journal, 59 (2013) 188–201.

[177] A.M. Nassimi, M. Jafari, H. Farrokhpour, M.H. Keshavarz, Constants of explosive limits, Chemical Engineering Science, 173 (2017) 384–389.

[178] M.G. Zabetakis, Flammability characteristics of combustible gases and vapors, in: Bureau of Mines Washington DC, 1965.

[179] T.A. Albahri, Prediction of the upper flammability limit of pure compounds, Asian Journal of Applied Science and Engineering, 5 (2016) 59–70.

[180] M. Al-Ghouti, Y. Al-Degs, F. Mustafa, Determination of hydrogen content, gross heat of combustion, and net heat of combustion of diesel fuel using FTIR spectroscopy and multivariate calibration, Fuel, 89 (2010) 193–201.

[181] T. Daubert, R. Danner, API technical data book-petroleum refining; American Petroleum Institute (API): Washington DC, 1997, There is no corresponding record for this reference.

[182] T.A. Albahri, Enhanced method for predicting the properties of light petroleum fractions, Fuel, 85 (2006) 748–754.

[183] T.A. Albahri, Molecularly explicit characterization model (MECM) for light petroleum fractions, Industrial & Engineering Chemistry Research, 44 (2005) 9286–9298.

[184] A. D4809-13, Standard test method for heat of combustion of liquid hydrocarbon fuels by bomb calorimeter (precision method), in: ASTM International, West Conshohocken, PA, 2013.

[185] A. D3338/D3338M-09, Standard test method for estimation of net heat of combustion of aviation fuels, in: ASTM International, West Conshohocken, PA, 2009.

[186] A. D1405/D1405M-08, Standard test method for estimation of net heat of combustion of aviation fuels, in: ASTM International, West Conshohocken, PA, 2013.

[187] D4529-01, Standard test method for estimation of net heat of combustion of aviation fuels, in: ASTM International, West Conshohocken, PA, 2011.

[188] A. D4868-00, Standard test method for estimation of net and gross heat of combustion of burner and diesel fuels, in: ASTM International, West Conshohocken, PA, 2010.

[189] A. D240-02, Standard test method for heat of combustion of liquid hydrocarbon fuels by bomb calorimeter, in: ASTM International, West Conshohocken, PA, 2002.

[190] R. Cardozo, Prediction of the enthalpy of combustion of organic compounds, AIChE Journal, 32 (1986) 844–848.

[191] W.H. Seaton, B.K. Harrison, A new general method for estimation of heats of combustion for hazard evaluation, Journal of Loss Prevention in the Process Industries, 3 (1990) 311–320.

[192] D. Van Krevlen, Thermochemical properties: calculation of the free enthalpy of reaction from group contributions, Properties of Polymers, (1990) 629–639.

[193] R.N. Walters, Molar group contributions to the heat of combustion, Fire and Materials, 26 (2002) 131–145.

[194] F.Y. Hshieh, D.B. Hirsch, H.D. Beeson, Ignition and combustion of low-density polyimide foam, Fire and Materials, 27 (2003) 119–130.

[195] F. Gharagheizi, A simple equation for prediction of net heat of combustion of pure chemicals, Chemometrics and Intelligent Laboratory Systems, 91 (2008) 177–180.

[196] H. Cao, J. Jiang, Y. Pan, R. Wang, Y. Cui, Prediction of the net heat of combustion of organic compounds based on atom-type electrotopological state indices, Journal of Loss Prevention in the Process Industries, 22 (2009) 222–227.

[197] M.H. Keshavarz, B.E. Saatluo, A. Hassanzadeh, A new method for predicting the heats of combustion of polynitro arene, polynitro heteroarene, acyclic and cyclic nitramine, nitrate ester and nitroaliphatic compounds, Journal of Hazardous Materials, 185 (2011) 1086–1106.

[198] Y. Pan, J. Jiang, R. Wang, J. Jiang, Predicting the net heat of combustion of organic compounds from molecular structures based on ant colony optimization, Journal of Loss Prevention in the Process Industries, 24 (2011) 85–89.

[199] F. Gharagheizi, S.A. Mirkhani, A.-R. Tofangchi Mahyari, Prediction of standard enthalpy of combustion of pure compounds using a very accurate group-contribution-based method, Energy & Fuels, 25 (2011) 2651–2654.

[200] A.-O. Diallo, G. Fayet, C. Len, G. Marlair, Evaluation of heats of combustion of ionic liquids through use of existing and purpose-built models, Industrial & Engineering Chemistry Research, 51 (2012) 3149–3156.

[201] H.Y. Cao, R. Wang, A new method for predicting the net heat of combustion of organic compounds, in: Advanced Materials Research, Trans Tech Publ, 2013, pp. 210–215.

[202] T.A. Albahri, Method for predicting the standard net heat of combustion for pure hydrocarbons from their molecular structure, Energy Conversion and Management, 76 (2013) 1143–1149.

[203] K. Zarei, M. Atabati, S. Moghaddary, Predicting the heats of combustion of polynitro arene, polynitro heteroarene, acyclic and cyclic nitramine, nitrate ester and nitroaliphatic compounds using bee algorithm and adaptive neuro-fuzzy inference system, Chemometrics and Intelligent Laboratory Systems, 128 (2013) 37–48.

[204] T.A. Albahri, Accurate prediction of the standard net heat of combustion from molecular structure, Journal of Loss Prevention in the Process Industries, 32 (2014) 377–386.

[205] A. Salmon, D. Dalmazzone, Prediction of enthalpy of formation in the solid state (at 298.15 K) using second-order group contributions. Part 1. Carbon-hydrogen and carbon-hydrogen-oxygen compounds, Journal of Physical and Chemical Reference Data, 35 (2006) 1443–1457.

[206] A. Salmon, D. Dalmazzone, Prediction of enthalpy of formation in the solid state (at 298.15 K) using second-order group contributions – Part 2: Carbon-hydrogen, carbon-hydrogen-oxygen, and carbon-hydrogen-nitrogen-oxygen compounds, Journal of Physical and Chemical Reference Data, 36 (2007) 19–58.

[207] B.M. Rice, S.V. Pai, J. Hare, Predicting heats of formation of energetic materials using quantum mechanical calculations, Combustion and Flame, 118 (1999) 445–458.

[208] E.F. Byrd, B.M. Rice, Improved prediction of heats of formation of energetic materials using quantum mechanical calculations, The Journal of Physical Chemistry A, 110 (2006) 1005–1013.

[209] S. Kondo, A. Takahashi, K. Tokuhashi, Theoretical calculation of heat of formation and heat of combustion for several flammable gases, Journal of Hazardous Materials, 94 (2002) 37–45.

[210] D. Saldana, L. Starck, P. Mougin, B. Rousseau, B. Creton, On the rational formulation of alternative fuels: melting point and net heat of combustion predictions for fuel compounds using machine learning methods, SAR and QSAR in Environmental Research, 24 (2013) 259–277.

[211] E. Sagadeev, V. Barabanov, Calculations of the enthalpies of combustion of organic compounds by the additive scheme, Russian Journal of Physical Chemistry A, Focus on Chemistry, 80 (2006) S152–S162.

[212] E. Sagadeev, A. Gimadeev, V. Barabanov, Calculation of the heat of combustion for organonitrogen compounds using a group additivity scheme, Theoretical Foundations of Chemical Engineering, 43 (2009) 108–118.

[213] M.H. Keshavarz, M. Oftadeh, New method for estimating the heat of formation of CHNO explosives in crystalline state, High Temperatures-High Pressures, 35 (2004) 499.

[214] M.H. Keshavarz, H. Sadeghi, A new approach to predict the condensed phase heat of formation in acyclic and cyclic nitramines, nitrate esters and nitroaliphatic energetic compounds, Journal of Hazardous Materials, 171 (2009) 140–146.

[215] M.H. Keshavarz, Theoretical prediction of condensed phase heat of formation of nitramines, nitrate esters, nitroaliphatics and related energetic compounds, Journal of Hazardous Materials, 136 (2006) 145–150.

[216] M.H. Keshavarz, M.K. Tehrani, H. Pouretedal, A. Semnani, New pathway for quick estimation of gas phase heat of formation of non-aromatic energetic compounds, Indian Journal of Engineering & Materials Sciences, 13 (2006) 542–548.

[217] M.H. Keshavarz, Prediction of the condensed phase heat of formation of energetic compounds, Journal of Hazardous Materials, 190 (2011) 330–344.

[218] M.H. Keshavarz, Predicting condensed phase heat of formation of nitroaromatic compounds, Journal of Hazardous Materials, 169 (2009) 890–900.

[219] M.H. Keshavarz, A simple procedure for calculating condensed phase heat of formation of nitroaromatic energetic materials, Journal of Hazardous Materials, 136 (2006) 425–431.

[220] M.H. Keshavarz, M.K. Tehrani, A new method for determining gas phase heat of formation of aromatic energetic compounds, Propellants, Explosives, Pyrotechnics, 32 (2007) 155–159.

[221] M.H. Keshavarz, M. Zamani, F. Atabaki, K.H. Monjezi, Reliable approach for prediction of heats of formation of polycyclic saturated hydrocarbons using recently developed density functionals, Computational and Theoretical Chemistry, 1011 (2013) 30–36.

[222] M.H. Keshavarz, H.R. Pouretedal, A.R. Ghaedsharafi, S.E. Taghizadeh, Simple method for prediction of the standard Gibbs free energy of formation of energetic compounds, Propellants, Explosives, Pyrotechnics, 39 (2014) 815–818.

[223] M. Oftadeh, M.H. Keshavarz, R. Khodadadi, Prediction of the condensed phase enthalpy of formation of nitroaromatic compounds using the estimated gas phase enthalpies of formation by the PM3 and B3LYP methods, Central European Journal of Energetic Materials, 11 (2014) 143–156.

[224] B. Nazari, M.H. Keshavarz, M. Hamadanian, S. Mosavi, A.R. Ghaedsharafi, H.R. Pouretedal, Reliable prediction of the condensed (solid or liquid) phase enthalpy of formation of organic energetic materials at 298 K through their molecular structures, Fluid Phase Equilibria, 408 (2016) 248–258.

[225] M. Jafari, M.H. Keshavarz, Simple approach for predicting the heats of formation of high nitrogen content materials, Fluid Phase Equilibria, 415 (2016) 166–175.

[226] M. Jafari, M.H. Keshavarz, M.R. Noorbala, M. Kamalvand, A reliable method for prediction of the condensed phase enthalpy of formation of high nitrogen content materials through their gas phase information, ChemistrySelect, 1 (2016) 5286–5296.

[227] M.H. Keshavarz, T. M. Klapotke, The properties of energetic materials, Walter de Gruyter GmbH & Co KG, 2017.

[228] P.E. Rouse Jr, Enthalpies of formation and calculated detonation properties of some thermally stable explosives, Journal of Chemical and Engineering Data, 21 (1976) 16–20.

[229] R. Lyon, R. Walters, S. Stoliarov, Thermal analysis of flammability, Journal of Thermal Analysis and Calorimetry, 89 (2007) 441.

[230] H. Lu, C.A. Wilkie, Synergistic effect of carbon nanotubes and decabromodiphenyl oxide/Sb 2 O 3 in improving the flame retardancy of polystyrene, Polymer Degradation and Stability, 95 (2010) 564–571.

[231] R.E. Lyon, Solid-state thermochemistry of flaming combustion, in: A.F. Grand, C.A. Wilkie (Eds.), Fire retardancy of polymeric materials, Marcel Dekker, Inc., New York, 2000, pp. 391–447.

[232] R.E. Lyon, Heat Release Capacity, in: Proceedings of the Fire & Materials Conference, Interscience Communications Limited: London, England, San Francisco, CA, 2001, pp. 285–300.

[233] R.E. Lyon, Heat release kinetics, Fire and Materials, 24 (2000) 179–186.

[234] C. Huggett, Estimation of rate of heat release by means of oxygen consumption measurements, Fire and Materials, 4 (1980) 61–65.

[235] R.N. Walters, S.M. Hackett, R.E. Lyon, Heats of combustion of high temperature polymers, Fire and Materials, 24 (2000) 245–252.

[236] R.N. Walters, R.E. Lyon, Molar group contributions to polymer flammability, Journal of Applied Polymer Science, 87 (2003) 548–563.

[237] H. Zhang, R.J. Farris, P.R. Westmoreland, Low flammability and thermal decomposition behavior of poly (3, 3'-dihydroxybiphenylisophthalamide) and its derivatives, Macromolecules, 36 (2003) 3944–3954.

[238] B.-Y. Ryu, S. Moon, I. Kosif, T. Ranganathan, R.J. Farris, T. Emrick, Deoxybenzoin-based epoxy resins, Polymer, 50 (2009) 767–774.

[239] H. Zhang, P. Westmoreland, R. Farris, E. Coughlin, A. Plichta, Z. Brzozowski, Thermal decomposition and flammability of fire-resistant, UV/visible-sensitive polyarylates, copolymers and blends, Polymer, 43 (2002) 5463–5472.

[240] A.A. Mir, S. Wagner, R.H. Krämer, P. Deglmann, T. Emrick, Deoxybenzoin-containing polysulfones and polysulfoxides: Synthesis and thermal properties, Polymer, 84 (2016) 59–64.

[241] C.A. Harper, Handbook of building materials for fire protection, McGraw-Hill, New York, 2004.

[242] R.E. Lyon, M.T. Takemori, N. Safronava, S.I. Stoliarov, R.N. Walters, A molecular basis for polymer flammability, Polymer, 50 (2009) 2608–2617.

[243] P.V. Parandekar, A.R. Browning, O. Prakash, Modeling the flammability characteristics of polymers using quantitative structure–property relationships (QSPR), Polymer Engineering & Science, 55 (2015) 1553–1559.

[244] M.H. Keshavarz, A. Dashtizadeh, H. Motamedoshariati, H. Soury, A simple model for reliable prediction of the specific heat release capacity of polymers as an important characteristic of their flammability, Journal of Thermal Analysis and Calorimetry, 128 (2017) 417–426.

[245] F. Atabaki, M.H. Keshavarz, N.N. Bastam, The simplest model for reliable prediction of the total heat release of polymers for assessment of their combustion properties, Journal of Thermal Analysis and Calorimetry, 131 (2018) 2235–2242.

[246] R.S. Jessup, Heats of combustion of diamond and graphite, Journal of Research of the National Bureau of Standards, 21 (1938) 475–490.

[247] W. Wang, L. Perng, G. Hsiue, F. Chang, Characterization and properties of new silicone-containing epoxy resin, Polymer, 41 (2000) 6113–6122.

[248] S. Liu, X. Lang, H. Ye, S. Zhang, J. Zhao, Preparation and characterization of copolymerized aminopropyl/phenylsilsesquioxane microparticles, European Polymer Journal, 41 (2005) 996–1001.

[249] H. Wang, X.F. Wang, C.L. Yu, Preparation and Characterization of Copolymerized Aminohexylaminomethyl/Phenylsilsesquioxane Microparticles, in: Applied Mechanics and Materials, Trans Tech Publ, 2012, pp. 114–117.

[250] C.-L. Chiang, C.-C.M. Ma, Synthesis, characterization, thermal properties and flame retardance of novel phenolic resin/silica nanocomposites, Polymer Degradation and Stability, 83 (2004) 207–214.

[251] X. Qian, L. Song, Y. Bihe, B. Yu, Y. Shi, Y. Hu, R.K. Yuen, Organic/inorganic flame retardants containing phosphorus, nitrogen and silicon: preparation and their performance on the flame retardancy of epoxy resins as a novel intumescent flame retardant system, Materials Chemistry and Physics, 143 (2014) 1243–1252.

[252] F. Atabaki, M.H. Keshavarz, N. Noorollahy Bastam, A simple method for the reliable prediction of char yield of polymers, Zeitschrift für anorganische und allgemeine Chemie, 643 (2017) 1049–1056.

[253] S. Gagarin, A. Gyul'maliev, Heat of vaporization of methyl-substituted phenol derivatives, Coke and Chemistry, 50 (2007) 232–236.

[254] M.H. Keshavarz, Estimation of the flash points of saturated and unsaturated hydrocarbons, Indian Journal of Engineering & Materials Sciences, 19 (2012) 269–278.

[255] D.R. Lide, W.M. Hynes, CRC Handbook of Chemistry and Physics, 85th ed., CRC Press, Boca Raton, Florida, 2010.

[256] C.L. Yaws, Yaws' Handbook of thermodynamic and physical properties of chemical compounds, Knovel, New York, 2003.

[257] Q. Jia, Q. Wang, P. Ma, S. Xia, F. Yan, H. Tang, Prediction of the flash point temperature of organic compounds with the positional distributive contribution method, Journal of Chemical & Engineering Data, 57 (2012) 3357–3367.

[258] G.-B. Wang, C.-C. Chen, H.-J. Liaw, Y.-J. Tsai, Prediction of flash points of organosilicon compounds by structure group contribution approach, Industrial & Engineering Chemistry Research, 50 (2011) 12790–12796.

[259] G.W. Gokel, Dean's Hanbook of organic chemistry, McGraw-Hill, 2004.

[260] J. Bond, Sources of ignition: Flammability characteristics of chemicals and products, Elsevier, 2017.

[261] J. Harris, Autoignition temperatures of military high explosives by differential thermal analysis, Thermochimica Acta, 14 (1976) 183–199.

[262] R.J. Lewis, Hazardous chemicals desk reference, John Wiley & Sons, Hoboken, 2008.

[263] R.P. Pohanish, Sittig's handbook of toxic and hazardous chemicals and carcinogens, William Andrew, Kidlington, Oxford OX5 1GB, UK, 2017.

[264] C.L. Yaws, C. Gabbula, Yaws' handbook of thermodynamic and physical properties of chemical compounds, Knovel, Norwich, 13 Eaton Ave, Norwich NY 13815 United Statesa, 2003.

[265] R.A. Lewis, Hawley's condensed chemical dictionary, John Wiley & Sons, Hoboken, 2016.

[266] P. Patnaik, A comprehensive guide to the hazardous properties of chemical substances, John Wiley & Sons, Hoboken, 2007.

[267] R.E. Lyon, T. Emrick, Non-halogen fire resistant plastics for aircraft interiors, Polymers for Advanced Technologies, 19 (2008) 609–619.

About the Author

Mohammad Hossein Keshavarz (b. 1966), received a BSc in chemistry in 1988 from Shiraz University, Iran. He also received an MSc and a PhD at Shiraz University in 1991 and 1995, respectively. From 1997 until 2008, he was Assistant Professor, Associate Professor, and Professor of Physical Chemistry at the University of Malek Ashtar in Shahin Shahr, Iran. Since 1997, he has been Lecturer and researcher at the Malek Ashtar University of Technology, Iran. He is also the editor of two research journals in the Persian language. Keshavarz has published over 300 scientific papers in international peer-reviewed journals, three book chapters, and six books in the field of assessment of energetic materials (four books in Persian and two books in English).

Index

https://doi.org/10.1515/9783110572223-011

.

www.ingramcontent.com/pod-product-compliance
Lightning Source LLC
Chambersburg PA
CBHW061413210326
41598CB00035B/6198